MINISTRY OF AGRICULTURE, FISHERIES AND FOOD

LEAD IN FOOD: PROGRESS REPORT

*The twenty-seventh report of the
Steering Group on Food Surveillance
The Working Party on Inorganic
Contaminants in Food
Third Supplementary Report on Lead*

Food Surveillance Paper No. 27

LONDON

HER MAJESTY'S STATIONERY OFFICE

ISBN 0 11 242886 X

Previous Food Surveillance Papers published by HMSO, are as follows:

Food Surveillance Paper No.	Title
1	The surveillance of food contamination in the United Kingdom
2	Survey of vinyl chloride content of polyvinyl chloride for food contact and of foods
3	Survey of vinylidene chloride levels in food contact materials and in foods
4	Survey of mycotoxins in the United Kingdom
5	Survey of copper and zinc in food
6	Survey of acrylonitrile and methacrylonitrile levels in food contact materials and in foods
7	Survey of dieldrin residues in food
8	Survey of arsenic in food
9	Report of the Working Party on Pesticide Residues (1977–1981)
10	Survey of lead in food: second supplementary report
11	Survey of styrene levels in food contact materials and in foods
12	Survey of cadmium in food: first supplementary report
13	Polychlorinated biphenyl (PCB) residues in food and human tissues
14	Steering Group on Food Surveillance progress report 1984
15	Survey of aluminium, antimony, chromium, cobalt, indium, nickel, thallium and tin in food
16	Report of the Working Party on Pesticide Residues (1982 to 1985)
17	Survey of mercury in food: second supplementary report
18	Mycotoxins
19	Survey of colour usage in food
20	Nitrate, nitrite and N-nitroso compounds in food
21	Survey of plasticiser levels in food contact materials and in foods
22	Anabolic, anthelmintic and antimicrobial agents
23	The British diet: finding the facts
24	Food Surveillance 1985 to 1988
25	Report of the Working Party on Pesticide Residues: 1985–88
26	Progress Report of the Working Party on Chemical Contaminants from Food Contact Materials for 1984 to 1988

Earlier reports of the Working Party on the Monitoring of Foodstuffs for Heavy Metals published by HMSO prior to the Food Surveillance Paper series are as follows:

Survey of Mercury in Food	1971
Survey of Lead in Food	1972
Survey of Mercury in Food: A Supplementary Report	1973
Survey of Cadmium in Food	1973
Survey of Lead in Food: First Supplementary Report	1975

STEERING GROUP ON FOOD SURVEILLANCE

This report was considered and recommended for publication by the Steering Group on Food Surveillance whose membership was then as follows:

Dr R N Crossett (Chairman)	BSc, BAgr, DPhil, FIFST	Ministry of Agriculture, Fisheries and Food
Dr M E Knowles (Deputy Chairman)	BPharm, PhD, CChem, FRSC, FIFST	Ministry of Agriculture, Fisheries and Food
Dr D H Watson (Secretary)	BSc, PhD, CBiol, MIBiol	Ministry of Agriculture, Fisheries and Food
Mr A P Patel (Administrative Secretary)		Ministry of Agriculture, Fisheries and Food
Mrs E A J Attridge		Ministry of Agriculture, Fisheries and Food
Professor D Barltrop	MD, BSc, FRCP, DCH	University of London
Dr P J Bunyan	BSc, PhD, DSc, CChem, FRSC, CBiol, FIBiol, FIFST	Ministry of Agriculture, Fisheries and Food
Dr G I Forbes	FRACMA, FACOM, FREHIS, MFCM, LRCS+P, DPH, DIH, DTM+H, DMSA	Scottish Home and Health Department
Mr A J Harrison	MBE, MChemA, CChem, FRSC, FIFST, FRSH	Public Analyst
Dr N J King	BSc, PhD	Department of the Environment
Professor P J Lawther	CBE, MB, BS, DSc, FRCP, FFOM	Consultant
Dr I F H Purchase	BVSc, PhD, MRCVS, FRCPath, CBiol, FIBiol	ICI plc
Dr J H Steadman	MB, BS, MSc	Department of Health
Mr R S Stewart		Department of Agriculture and Fisheries for Scotland
Mr L G Weir		Department of Health
Dr J J Wren	MA, PhD, MFC, CChem, FRSC, FIFST, MHCIMA	Consultant

STEERING GROUP ON FOOD SURVEILLANCE

Terms of Reference

To keep under review the possibilities of contamination of any part of the national food supply, to review where necessary the intake of individual additives and nutrients and to recommend to Ministers responsible for food quality and safety the programme of work necessary to ensure that the food intake of the population is both safe and nutritious. To appoint expert Working Parties, acting according to such instructions as the Steering Group may give, to carry out specialist parts of the programme of work.

To consider reports made by Working Parties and to decide what action, including consultation with the Government's advisory committees and other bodies having an interest in the subject matter or the implications of the reports, should be recommended.

To submit the findings of the Working Party reports, where appropriate, to the Ministers with appropriate recommendations as to publication.

WORKING PARTY ON INORGANIC CONTAMINANTS IN FOOD

This report was prepared by the Working Party on Inorganic Contaminants in Food whose membership was then as follows:

Dr D J McWeeny (Chairman)	DSc, FIFST	Food Science Laboratory, Ministry of Agriculture, Fisheries and Food (MAFF)
Mr T J Coomes (until 1988)	BSc, FIFST	MAFF
Dr J L Facer (until 1988)	MB, BS, FRCP, MFCM	Department of Health and Social Security
Dr G I Forbes	FRACMA, FACOM, FREHIS, MFCM, LRCS+P, DPH, DIH, DTM+H, DMSA	Scottish Home and Health Department
Mr A Franklin	BSc	Fisheries Research Directorate, MAFF
Dr A M George	BSc, PhD	Welsh Office
Mr A J Harrison	MBE, MChemA, CChem, FRSC, FIFST, FRSH	Public Analyst
Dr J S Hislop	BSc, PhD	Atomic Energy Research Establishment, Harwell
Mr G Kirby	LRSC	Food and Drink Federation
Mr I Lumley	MSc, CChem, MRSC	Laboratory of the Government Chemist
Dr R C Massey	BSc, PhD	Food Science Laboratory, MAFF
Mrs E Owen		MAFF
Mr J F A Thomas	BSc	Department of the Environment
Dr G Topping	BSc, PhD, CChem, FRSC	Marine Laboratory (Aberdeen) Department of Agriculture and Fisheries for Scotland
Mr R J Unwin	BSc	MAFF
Dr M Waring (from 1988)	MA, MB, FRCS	Department of Health

Secretariat:

Scientific:

Miss G A Smart	MSc, MIInf Sci	MAFF
Dr J C Sherlock (until 1987)	BSc, PhD, CChem, FRSC, FIFST	MAFF
Dr C E Fisher (from 1987 to 1988)	MA, DPhil	MAFF
Dr D H Watson (from 1988 to 1989)	BSc, PhD, CBiol, MIBiol	MAFF
Dr J A Norman (until 1987)	BSc, DPhil	MAFF

Administrative:

Mrs R F Bott	MAFF

WORKING PARTY ON INORGANIC CONTAMINANTS IN FOOD

Terms of Reference

To determine the amounts of certain heavy metals and other inorganic substances in food in the United Kingdom and to make reports. The results of the surveys to be submitted to the Committee on Toxicity of Chemicals in Food, Consumer Products and the Environment and the Food Advisory Committee who would inform the Ministers if any hazard to health existed.

CONTENTS

GLOSSARY OF TERMS, ABBREVIATIONS AND UNITS

Provisional Tolerable Weekly Intake (PTWI): The Joint FAO/WHO Expert Committee on Food Additives[1] adopted the following approach. Since heavy metal contaminants are able to accumulate within the body at a rate and to an extent determined by the level of intake and by the chemical form of the heavy metal present in food, the basis on which intake is expressed should be more than the amount corresponding to a single day. In addition, individual foods may contain above-average levels of a heavy metal contaminant, so that consumption of such foods on any particular day greatly enhances that day's intake. Accordingly, the Provisional Tolerable Intake was expressed on a weekly basis.

The term 'tolerable' signifies permissibility rather than acceptability. The use of the term 'provisional' expresses the tentative nature of the evaluation, in view of the paucity of reliable data on the consequences of human exposure at levels approaching those with which the Committee was concerned.

Abbreviations used in this report:

AERE: Atomic Energy Research Establishment
AFRC: Agricultural and Food Research Council
EC: European Commission
FAO: Food and Agriculture Organization
LOD: Limit of Determination
RCEP: Royal Commission on Environmental Pollution
WHO: World Health Organization

Summary of units used in this report

mg, milligram:	one thousandth (10^{-3}) of a gram
μg, microgram:	one millionth (10^{-6}) of a gram
kg, kilogram:	one thousand (10^{3}) grams
mg/kg:	milligrams per kilogram, equivalent to parts in 10^{6} by weight
ml, millilitre:	one thousandth (10^{-3}) of a litre
l:	litre
g/l:	grams per litre
mg/l:	milligrams per litre
μg/l:	micrograms per litre
μg/100 ml:	micrograms per hundred millilitres
Bq:	Becquerel
μg/m^{3}:	micrograms per metre cubed

SUMMARY

This report describes surveys and research which have been carried out since the previous report on Lead in Food was published in 1982[2]. As well as giving estimates of lead intakes in the United Kingdom and presenting data on lead levels in individual foods, consideration is given to the pathways by which lead reaches food and to the absorption and excretion of lead by humans. Results from the British Total Diet Study indicate that during the period 1982–1987 average lead intakes over the whole population were between 0.02 and 0.07 mg per day. This excludes the contribution to lead intakes from drinking water. Duplicate diet studies have been carried out for both adults and children resident in an area where exposure to lead was known to be low. The mean dietary lead intake of women was estimated as 0.31 mg per week and the mean lead intake of the children as 0.11 mg per week. The mean lead concentrations of dietary samples from the 2 different types of study were similar. However, in areas with elevated levels of lead in tap water, estimated lead intakes of both adults and children were found to be higher and in some cases the Provisional Tolerable Weekly Intake recommended by the Food and Agriculture Organisation/ World Health Organisation (FAO/WHO) of 3 mg per week was exceeded. For regular consumers of alcoholic beverages, beer and wine may make a significant contribution to dietary lead intake and this is reflected in the higher blood lead levels found for drinkers compared to non-drinkers. It has also been found that lead in beer is absorbed more readily (20 per cent) from the gastrointestinal tract than lead from the rest of the diet (less than 10 per cent). The contribution from the deposition of airborne lead on soil and crops to lead in whole diets has been estimated to be between 13 per cent and 31 per cent for children. For individual food plants, however, a much higher percentage of the lead content may derive from aerial deposition (40–100 per cent). Where crops are contaminated by lead from the soil, much of this may be removed by normal culinary preparation.

It is the Government's policy to ensure that every practicable measure is taken to reduce human exposure to lead. In this respect, the Ministry has responsibility for controlling the levels of lead in food and beverages excluding drinking water, which is the responsibility of Environment Departments. Since 1982 action has been taken on a number of fronts:

— the amount of lead in petrol has been reduced;

— changes in can-making technology have resulted in the replacement of lead-soldered cans by using welded side-seams;

— contamination of draught beer by lead has been reduced by replacing dispensing equipment made from brass and other lead-containing alloys;

— action has been initiated to phase out the use of lead capsules to seal wine bottles.

1

Programmes of work have been carried out to monitor the effects of these changes on lead levels in food and the results are given in this report.

The views of the Committee on Toxicity of Chemicals in Food, Consumer Products and the Environment and the Food Advisory Committee are given in Appendices I and II respectively.

INTRODUCTION

1. The Working Party on the Monitoring of Foodstuffs for Heavy Metals and Other Inorganic Contaminants, was set up in 1971, and has recently been re-named the Working Party on Inorganic Contaminants in Food. The Working Party has published the results of surveys of foods for the presence of mercury, lead, cadmium, arsenic, copper, zinc, aluminium, antimony, chromium, cobalt, indium, nickel, thallium and tin. Lead was first considered[3] in 1972 and since then 2 supplementary reports have been produced[2,4], the most recent being published in 1982. The results of these surveys were submitted for comment to the Food Advisory Committee (FAC)* and the Committee on Toxicity of Chemicals in Food, Consumer Products and the Environment (COT)†.

2. In their consideration of *Survey of Lead in Food: Second Supplementary Report*, the FAC and the COT made a number of recommendations[2] for future work:

— Monitoring of the lead content of canned food and of the national diet should continue.

— Lead intakes of critical groups such as children and pregnant women should be monitored.

— Work should be carried out on the routes by which lead reaches man.

— The factors affecting the absorption of lead derived from food should be investigated.

3. The Working Party has also been guided in its programme of research on lead by 2 other reports—the report *Lead in the Environment* produced by the Royal Commission on Environmental Pollution[5] and the Food Additives and Contaminants Committee (FACC) report *Review of Metals in Canned Food*[6]. These 2 reports are considered in more detail below.

Royal Commission on Environmental Pollution—Report on Lead

4. The Royal Commission on Environmental Pollution (RCEP) report on lead[4] was published in 1983. It contains 2 recommendations of relevance to food:

— Recommendation 18. There should be a continuing effort to gain a better understanding of the various pathways and mechanisms by which food is contaminated with lead.

— Recommendation 19. More data should be obtained on the lead content of alcoholic drinks at the point of consumption.

---

*Formerly the Food Additives and Contaminants Committee (FACC).
†Formerly the Toxicity Sub-Committee of the Committee on Medical Aspects of Chemicals in Food and the Environment (TSC).

One further recommendation (number 26), that of reducing the maximum permitted lead content of petrol from 0.40 g/l to 0.15 g/l, also has implications for the lead content of food.

5.   Work relating to recommendations 18 and 19 was already underway following consideration of the *Second Supplementary Report on Lead*[2] by Government Advisory Committees. In addition, new work was initiated to monitor the effect of reducing the concentration of lead in petrol on lead levels in food.

FACC Review of Metals in Canned Foods

6.   The FACC published a review of metals in canned foods[6] in 1983. The Committee noted that changes in can-making technology were taking place which should lead to a reduction in the levels of lead found in canned food, and accordingly recommended:

— that the lead content of canned foods in the diet should be regularly monitored so that the impact of changes in can-making technology on the lead content of canned food could be determined; and

— that statutory limits for lead in canned foods should be brought into line with the limits for the corresponding foods not contained in cans.

Work Commissioned

7.   In response to the recommendations made by the FACC, COT and RCEP, the Working Party commissioned work in the following areas:

— Monitoring of samples from the British Total Diet Study[7] and individual foods.

— Lead intakes by individuals (duplicate diet studies).

— Work to monitor the effects of the reduction of lead in petrol, and changes in can-making technology.

— Studies of pathways by which lead reaches food, for example atmospheric deposition and uptake by plants from soil.

— Studies on absorption and excretion of lead by humans.

Results from all these areas are discussed in this report, together with some from projects completed too late for inclusion in the *Second Supplementary Report on Lead*[2]. A full list of research projects is given in Appendix III.

8.   The total cost of the work described in this report was £690,000 which represented 40 per cent of the total spent by MAFF, under the guidance of the Working Party, on research on heavy metals in the diet between 1982 and 1988. During this period the relative proportion of expenditure for work on lead decreased from 71 per cent to 20 per cent, as the major gaps in knowledge about the effects of lead on humans and the environment have been filled. The money

was apportioned between the subject areas (para.7) and the analytical quality assurance schemes (paras.12–14) as shown in Appendix III.

Current Legislation

9.   *The Lead in Food Regulations 1961* laid down a general limit of 2 mg/kg in food, with exceptions for certain specified foods[8]. Similar legislation applies in Scotland and Northern Ireland. However the general limit for lead in food was decreased[9] to 1 mg/kg in *The Lead in Food Regulations 1979* and a further change was made by *The Lead in Food (Amendment) Regulations 1985* in which[10] the exceptional limit of 2 mg/kg for canned foods was also lowered to 1 mg/kg. The maximum content of lead in natural mineral waters is limited to 10 μg/l by *The Natural Mineral Waters Regulations 1985*[11].

10.   Legislation reducing the maximum permitted lead content of leaded petrol from 0.40 to 0.15 g/l came into effect[12] from 13 December 1985. It was intended to be an intermediate stage in the complete phasing out of lead additives in petrol.

Analytical Methods

11.   A wide variety of methods for the determination of lead has been used, with a concomitant variation in the limits of determination (LOD) achieved. Usually the choice of method is determined by the precision and accuracy required and this in turn determines the LOD. For the Total Diet Study, where changes in lead concentration from year to year are expected to be very small, the LODs are as low as possible (0.01–0.05 mg/kg). On the other hand when considering canned food, interest lay in samples containing relatively high amounts of lead so that the acceptable LOD was higher (0.1 mg/kg) to minimise analytical costs. In most cases atomic absorption spectrophotometry was used following wet oxidation of the sample[13].

Analytical Quality Assurance

12.   In the *Steering Group on Food Surveillance 1984 Progress Report*[14] the Steering Group stated that 'Wherever possible laboratories producing results from the analysis of chemicals in food should take part in external quality assurance programmes to be sure of the accuracy, reproducibility and precision of their methods'. International analytical intercomparison studies for lead, cadmium and other metals in foodstuffs[15, 16] had indicated that much of the analytical data was of poor quality. The Working Party therefore initiated a survey to assess the quality of results of analyses of food for lead and cadmium in the UK[17]. An Analytical Quality Assurance (AQA) Sub-Group was established by the Working Party to co-ordinate the survey and to establish criteria by which to assess the results.

B   5

13. The details of the initial intercomparison exercise are given in Appendix V. The criteria defined by the sub-group were intended to be relatively undemanding, but out of the 28 participating laboratories, only 4 achieved the criteria for 4 or more of the 8 samples supplied for both lead and cadmium[17].

14. The results of this exercise were discussed by the Working Party and the Steering Group on Food Surveillance. It was agreed that a further scheme was necessary and should be designed specifically with the object of bringing about an improvement in analytical performance. Approximately 80 laboratories were identified and approached to see whether they wished to participate in the scheme. Operation of the scheme was sub-contracted to AERE, Harwell who administered the scheme on behalf of the sub-group, produced the materials required and collected and evaluated the data. Details of this further main scheme are given in Appendix V. Forty-seven laboratories participated in all its stages. Based on the criteria set by the sub-group only one laboratory correctly analysed every sample, and only 3 laboratories correctly analysed 90 per cent or more samples. Most laboratories fell considerably short of the criteria. There have been many examples of failure to agree between laboratories, both nationally and internationally, in the collaborative testing of analytical methods for heavy metals in food. A further AQA scheme, which is partly self-financing has been initiated. It should be emphasised that the data presented in this report were produced by government laboratories and other laboratories that achieved a high level of performance in the AQA scheme.

RESULTS AND APPRAISAL

Dietary Intake of Lead from the British Total Diet Study

15. Table 1 of Appendix IV gives the lead concentrations in samples of foods obtained since 1982 from individual centres in the British Total Diet Study. The Total Diet is made up of food groups representing the average diet consumed in Great Britain by adults and children. Each group is prepared separately and then analysed for lead. The lead intake of the average person may then be estimated from a knowledge of the concentration of lead in each food group and the relative weight of each food group consumed each day.

16. Details of the composition of the food groups are given elsewhere[7]. The beverages group is effectively diluted prior to analysis since tea leaves are not incorporated directly but are extracted with hot distilled water, similar to normal culinary practice. The tea infusion is then added to the dried coffee, concentrated soft drinks and ready-to-drink soft drinks which make up the rest of the beverages group. This dilution is allowed for by correcting the concentrations so that the values given in Table 1 refer to undiluted beverages. Thus, dietary intakes as estimated from the Total Diet Study exclude the contribution made by lead in drinking water.

17.   After 1981, preparation of the Total Diet samples was carried out at the Agricultural and Food Research Council's (AFRC) Long Ashton Research Station. However due to changes in the AFRC in 1986, the site of preparation moved to the AFRC Institute of Food Research (Norwich Laboratory). Every effort was made to minimise the effects of the change. However as a check, duplicate samples of diets were purchased in 7 towns and one sample was prepared at the old site (Long Ashton) and one sample at the new (Norwich). There were no statistically significant differences in the concentrations of lead in any of the food groups of the 2 sets of diets. In summary, from 1981–85 diets were prepared at Long Ashton; from 1986 onwards the diets were prepared at Norwich. However, this change does not affect the validity of comparing data from 1986–87 with those from 1981–85.

18.   Lead concentrations in most food groups were at or below the limit of determination (LOD) [0.01 mg/kg for beverages; 0.02 mg/kg for potatoes; 0.05 mg/kg for other foods]. The exceptions were offal, which is known to accumulate lead, and the groups which contain canned food—fish, canned vegetables and fruit products. Since 1982 there has been a gradual decrease in the lead concentrations found in the 2 food groups which contain predominantly canned products, i.e. the canned vegetables and fruit products food groups. For example mean levels of lead in samples of the canned vegetables group fell from 0.17 mg/kg in 1982 to less than 0.05 mg/kg in 1987.

19.   Concentrations of lead in a food group of the British Total Diet Study are often less than the LOD. In these cases there are 2 commonly used methods of calculating intake:

   (i)   results quoted as less than the limit of determination may be taken as being equal to the limit of determination (upper bound value). This assumption produces an overestimate of intake; or

   (ii)   results quoted as less than the limit of determination are taken as being zero (lower bound value). This assumption produces an underestimate of intake.

The second method probably produces the answer which best reflects the actual situation but for the purposes of safety evaluations it is considered prudent to overestimate intakes rather than underestimate them; consequently intakes produced by method (i) are generally used. *Unless otherwise stated the tables reporting Total Diet data refer to intakes calculated by method (i).* Mean daily lead intakes (mg per person) for 1982–87 calculated using both methods are estimated as follows:

	1982	1983	1984	1985	1986	1987
Assuming values less than LOD equal that value	0.07	0.07	0.06	0.07	0.06	0.06
Assuming values less than LOD equal zero	0.04	0.02	0.02	0.03	0.02	0.02

20. These results show that estimated mean lead intakes for the UK for 1987 lay between 0.02 and 0.06 mg per day. Over the period 1982–87, estimated intakes averaged between 0.02 and 0.07 mg per day. Thus, although there has been a decrease in the lead concentration in some food groups over this period (para.18) this has not led to a significant decrease in the total average dietary intake since these food groups represent only 2–3 per cent by weight of the average diet.

21. Estimated intakes for the period 1982–87 were lower than those given in the previous report[2] for 1975–81 when the mean intake was estimated as 0.10 mg per day. This may be partly due to the re-organisation of the British Total Diet Study which took place in 1981. Prior to 1981, diets were prepared and cooked in a number of different locations using local tap water. After 1981, preparation of the diets was centralised and distilled water used rather than domestic tap water, thus eliminating the contribution made by lead in water to dietary intake.

22. The FAO/WHO Provisional Tolerable Weekly Intake for dietary lead[1] is 3 mg for an adult. The average person in Great Britain is estimated to have a dietary lead intake of up to 0.42 mg per week excluding the contribution from drinking water. This intake figure represents a maximum value as it was calculated by assuming that lead concentrations less than the LOD are equal to that value.

Individual Foodstuffs

23. Lead contents of individual food items are given in Table 2 of Appendix IV. As expected from previous surveys mean lead concentrations in most foods were very low (less than 0.05 mg/kg). Slightly higher levels were found in some samples of concentrated foods such as stock-cubes and chutneys, and also in some samples of soft fruit (both fresh and frozen). However lead levels in all the samples analysed were well below the statutory limit.

Selected Ethnic Foods

24. The results of analyses of a number of ethnic foodstuffs are given in Table 3 of Appendix IV. It can be seen that in contrast to the foodstuffs given in Table 2, levels were generally higher than the limit of determination, of the analytical method, of 0.05 mg/kg. The highest lead levels were found in one sample each of canned okra, canned anchovies, popadums and, in particular, Asian pickles. Two samples of Asian pickles and the sample of canned okra analysed exceeded the statutory limit for lead in food.

Vegetables

25. Data on the seasonal and geographic variation in the lead content of vegetables are given in Table 4 of Appendix IV. Samples were collected in each

of 4 areas of England during November 1982 and March 1983. The highest lead levels were found in samples of spinach (mean concentration 0.26 mg/kg). For other vegetables mean lead concentrations were generally less than 0.1 mg/kg. The ability of spinach to accumulate large amounts of lead has been noted in previous studies carried out by the Ministry[2, 18].

26. The mean lead content for all vegetables obtained in both months was greater in March than in November, consistent with previous literature reports that plants contain a higher level of lead in the winter months[2, 19]. No single location had vegetables with a consistently high lead content, although lead levels in the vegetables purchased in Dawlish tended to be lower than those purchased from the 3 urban locations. A study on the effects of the reduction of lead levels in petrol on the lead contents of retail green vegetables was carried out between 1983 and 1987. The vegetables chosen for study were curly kale, spring greens, lettuce and cabbage. The results showed that for some of the vegetables there was a significant decrease in lead concentration but for others there was no significant decrease. Details of this study and the conclusions are reported in Appendix VI.

Edible Bone Products

27. Following reports that ingestion of dietary supplements prepared from bonemeal had resulted in lead poisoning[20], a number of bone products available in the UK were analysed for lead in 1983. The results are given in Table 5 of Appendix IV. The bonemeal which had induced lead poisoning in USA contained from 60–190 mg/kg lead. In contrast the mean lead concentration in the bonemeal tablets analysed from this survey was 4.4 mg/kg and the range 1.3–9.4 mg/kg. Lead levels in the samples of other bone products were also relatively low.

Fish and Shellfish

28. Previous analyses of fish muscle have not detected lead at the limit of determination of 0.2 mg/kg wet weight in Scotland, England and Wales. Fish liver tissue has been considered as a possible alternative indicator tissue for the presence of contamination by lead in the marine environment, but in a comprehensive survey carried out in 1985 of fish and shellfish taken around the coasts of England and Wales, no fish liver samples were found which contained lead at concentrations above the present LOD of 0.6 mg/kg wet weight. There is therefore no indication of significant contamination of fish by lead.

29. Concentrations higher than the above mentioned determination limit can be found in the edible blue mussel (*Mytilus edulis*). Table 6 of Appendix IV lists levels found in samples taken during the 1985 survey. Considerable variation in lead levels was found but it should be noted that the survey included a number of industrial estuaries where mussels would not be likely to be taken for consumption. All levels however were well below the shellfish standard[9]

given in *The Lead in Food Regulations (1979)*, of 10 mg/kg wet weight, and none attained even the level that would prompt investigation, of 7 mg/kg wet weight.

Alcoholic Beverages

30.   In response to the recommendations of the Royal Commission on Environmental Pollution (para.4) a survey of lead levels in alcoholic beverages was undertaken in 1982[21]. Beer and lager hold the largest share of the UK alcoholic drinks market in terms of volume of liquid consumed, but wine is becoming increasingly popular. Traditionally tin-coated lead capsules are used to cap some wine bottles and this is an acknowledged source of lead contamination of wine as consumed[22, 23]. Thus wine was included in the monitoring programme.

31.   Approximately 90 per cent of canned and bottled beers contained 0.01 mg/l or less of lead, whereas nearly half the draught beers sampled contained more than 0.01 mg/l and 4 per cent contained more than 0.1 mg/l. Although none of the beer samples exceeded the statutory limit for lead in beer of 0.2 mg/l, it was considered prudent to trace the source of the contamination and attempt to eliminate it. Investigations showed that contamination arose through contact of the beer with various pieces of bronze and brass equipment used during its storage and dispense. The Brewers' Society therefore began a campaign to replace all bronze and brass equipment. To determine whether or not this action had been effective, a second survey was undertaken in 1985. Mean lead levels in both beer and lager had fallen, and in the second survey no samples contained more than 0.1 mg/l of lead. Eighty-three per cent of the samples contained 0.01 mg/l or less of lead compared to 55 per cent in the first survey. These results suggested that this action had had the desired effect.

32.   Wines were sampled directly from the bottle and then again after pouring, no attempt having been made to clean the mouth of the bottle. All the samples of unpoured wines contained less than 0.25 mg/l of lead, and the poured wines generally contained less than 0.3 mg/l of lead. However, five samples showed significant increases in lead content of up to 1.89 mg/l when poured. Four of these bottles showed obvious signs of lead corrosion products. *The Lead in Food Regulations*[9] set a statutory limit for lead in wine *as sold* of 1 mg/l. All of the bottles of wine examined complied with this limit, but 1 of the poured wine samples exceeded the statutory limit.

33.   In a second survey, 100 samples of poured wines (imported) were analysed for lead. All poured samples had been contained in bottles with lead closures and it was found that approximately 20 per cent contained more than 1 mg/l of lead, the statutory limit for wine as sold. A small survey of poured English wines was subsequently carried out, some of which were contained in bottles with lead closures. Six samples of wine in bottles without lead closures contained low lead concentrations ranging from 0.03–0.09 mg/l. Lead concentra-

tions from wine contained in bottles with lead closures (2 samples) were 0.07 mg/l and 0.7 mg/l. These data have been published[24] and a further survey of English wines is to be carried out. The data have also been brought to the attention of 2 independent expert UK committees, the Food Advisory Committee and the Committee on Toxicity of Chemicals in Food, Consumer Products and the Environment, who recommended that steps be taken to replace lead in the capsules with alternative materials, and that retailers and consumers be made aware of this potential source of lead exposure.

34.   In recent years use of tin-coated lead closures for sealing wine bottles has partially been replaced by the use of aluminium or plastic. However, since the use of lead capsules for bottles of quality wines still persists and as these wines may be stored for many years, it was considered that advice to wipe the tops and necks of bottles before pouring wine should be given. The Ministry therefore issued publicity in the form of a 'Food Facts' notice[25]. The wine trade are co-operating by including this advice in their own publicity material. The help of the catering trade in establishing this as good practice has also been sought.

Canned Foodstuffs

35.   It has already been noted that the British Total Diet Study food groups which contained the highest levels of lead were those which include canned foods or offal (para.18). The main source of lead in canned foods is the lead solder used to join the tinplate at the side seam on the outside. Modern methods of can-making dispense with the need for lead solder and the can-making industry has been switching to this technology over the last few years. In order to assess the effect on lead concentrations of changing from use of soldered to non-soldered cans, the Working Party set up a monitoring programme which lasted from 1983 to 1988.

36.   Seventeen types of canned food representing the varieties of canned food most commonly consumed in the UK (for example, spaghetti or baked beans) and including those most susceptible to contamination (for example, corned beef or sardines) were examined twice a year. The proportion of samples contained in non-soldered cans increased significantly during the survey period. At the beginning of the survey 12 per cent of samples were contained in non-soldered cans; in Spring 1985 this figure was 53 per cent and in Spring 1987 83 per cent of samples were contained in non-soldered cans. All samples contained in soldered cans in Spring 1987 were imported foods.

37.   Data on the lead concentrations found in individual foods in each phase of the survey are given in Table 7 of Appendix IV. Mean lead concentrations of less than 0.2 mg/kg were found in asparagus, tomato soup, baked beans, evaporated milk and cream throughout the survey period. For three of the most popular canned foods consumed in the UK, i.e. tomato soup, spaghetti and baked beans, lead levels decreased significantly during the survey period and are now less than 0.1 mg/kg. A decrease in mean lead concentration also

11

occurred between 1983 and 1987 for blackcurrants, pineapple, sardines, black-berries, fruit cocktail, red plums, cream and rhubarb.

38.   The foods which generally contained high levels of lead were sardines and corned beef. During 2 phases of the survey (Autumn 1985 and Spring 1986) extra sardine samples were analysed. In Autumn 1985 all of the samples were contained in soldered cans. The mean lead concentration was 0.99 mg/kg. In Spring 1986 all samples except one were canned in either folded aluminium or folded tinplate, and the mean lead concentration was 0.17 mg/kg. However, in Spring 1987 where 69 per cent of samples were contained in soldered cans the mean lead content was lower (0.13 mg/kg) and was very similar for samples contained in non-soldered cans (0.15 mg/kg) to that for samples contained in soldered cans (0.12 mg/kg). This indicates that lead levels were lower in this case owing to factors other than decreased use of can solder.

39.   There is evidence that once a can has been opened, the presence of oxygen accelerates the dissolution of lead from solder[26, 27]. Thus storage of food in opened soldered cans is likely to result in increased concentrations of lead in the food. A survey was carried out to investigate storage of both opened and unopened cans of food in the home[28]. The age of the can was ascertained wherever possible from the date code on the can end. Twelve per cent of the unopened cans for which information was available were found to be over 2 years old, 5 per cent were more than 3 years old and 3 per cent more than 4 years old. In 20 per cent of the households in the survey, food was found to be stored in opened cans: 30 per cent had been stored for more than 2 days after opening and nearly 10 per cent had been stored for more than 7 days.

40.   The Food Additives and Contaminants Committee recommended[6] that the public should be warned against storage of food in opened cans. In view of the widespread occurrence of this practice, the Ministry has conducted a public information campaign to educate people about the correct storage of canned food. Short items have been broadcast on the radio, and the Ministry has co-operated with women's magazines in producing articles giving advice on the use and storage of canned foods[29]; further publicity on this subject will follow.

Effect of Preparation and Cooking on the Lead Content of Vegetables

41.   As mentioned in previous reports[2, 4], up to 50 per cent of lead deposited from the atmosphere can be removed by washing vegetables and by discarding the outer leaves of leafy vegetables. A study was carried out recently on allot-ment vegetables grown in an area where air lead concentrations were higher than average. The results are given in Table 8 of Appendix IV. Samples of five types of vegetables (cabbage, carrots, leeks, lettuce, sprouts) were analysed after trimming, washing and cutting, and cooking. Washing and preparation (excluding cooking) were found to remove between 32 and 98 per cent of the

12

lead depending on the vegetable. Cooking in most cases removed further quantities of lead, so that after cooking the following proportions of lead had been removed: cabbage (54–94%), carrots (96–99%), leeks (76–99%), lettuce (42–97%), sprouts (42–94%).

42.   Although, in general, cooking leads to a reduction in the lead content of vegetables, in areas where the lead concentrations in water are higher than average, cooking water can be a significant source of lead intake. As described in our previous report, calculations indicate that at average concentrations of lead in water, of 0.02 mg/l, the contribution made by lead in water used for the preparation of foods and beverages to the total dietary lead intake is about 10 per cent, but this can rise to about 40 per cent as lead concentrations in water approach 0.1 mg/l[30].

Duplicate Diet Studies

43.   The Working Party has carried out four duplicate diet studies in order to estimate individuals' lead intakes. These data have been used to assess the number of people who exceed the FAO/WHO Provisional Tolerable Weekly Intake (PTWI) for lead and also to investigate relationships between dietary lead, water lead and blood lead. Emphasis has been placed on collection of data relating to children's lead intakes. The first study described (para.44) relates to a community exposed to elevated levels of lead in drinking water. The other studies (paras.45–52) were carried out in areas where water lead concentrations were relatively low. In duplicate diet studies, since whole diets are analysed, the contribution from lead in drinking water is included in the intake figure.

44.   In 1980 high lead concentrations were found in tap water in houses with lead plumbing in Ayr, Scotland. Dietary lead intakes were estimated and lead concentrations in blood and water were measured. Their relationships were investigated for adults and infants living in the area[31]. Estimated lead intakes of adults ranged between 0.2 and 11.9 mg per week, with approximately one third of the adults exceeding the relevant PTWI of 3 mg recommended by FAO/WHO. The infants' lead intakes ranged from 0.6 to 6.0 mg per week. Treatment of the water supply to reduce plumbosolvency was initiated in 1981. A follow-up study of lead concentrations in water and blood from the same population showed that these levels were within normal limits[32].

45.   In 1982 a control duplicate diet study was carried out in a rural area in north-east England. The study population was drawn from an area where exposure to lead was known to be minimal. Weights of food and drink items consumed were recorded in a diary and duplicate samples of all the food and drink consumed in a week were collected. Data on the amount of canned food and drink consumed were also collected. The study population consisted of women and pre-school age children.

46.   The mean lead intake of the children was 0.11 mg per week. There was no correlation between canned food consumption and dietary lead concentration. In contrast for the women there was a significant correlation between canned food consumption and dietary lead concentration. This difference between the women and children may have been due to the larger food intakes of the women and the consequently greater degree of certainty in measurement of canned food intakes and lead intakes. The women's mean lead intake was estimated as 0.31 mg per week. Estimated mean lead intake for the whole population from the British Total Diet Study in 1982 was 0.07 mg per day (0.49 mg per week; upper bound) and 0.04 mg per day (0.28 mg per week; lower bound). It has often been observed that intakes estimated from duplicate diet studies are lower than those derived from total diet studies. This may partly be because the upper bound figure of a total diet study is inevitably an overestimate. However, there is often a reluctance on the part of participants in duplicate diet studies to waste "good" food and they may also try to minimise the effort involved in participating in the study by eating less than normal[33]

47.   The mean lead concentration of the women's diets was 0.02 mg/kg. This figure is comparable to that calculated from the British Total Diet Study (e.g. 0.03 mg/kg, in 1982). Thus, for populations not exposed to a local source of lead, the metal concentrations measured in Total Diet samples and hence the intakes estimated from them provide a reasonable estimation of the lead intake of the average person (assuming that the consumption data on which the calculations are based are reliable).

48. Several studies[34, 35] had indicated that children in some groups belonging to Asian cultures have higher blood lead concentrations than others. To investigate this, a duplicate diet study was carried out on pre-school age children living in Harrow, London[36]. The children were sub-divided into 3 study groups: Caucasian children, Asian vegetarian children and Asian non-vegetarian children. Lead intakes were determined via a duplicate diet study of one week's duration and information on the child's hand washing habits was gathered using a questionnaire. Blood samples were analysed for lead and the 25-hydroxy metabolite of vitamin D.

49.   The mean lead intake of Asian children (both vegetarian and non-vegetarian) was estimated at 0.11 mg per week, and that of the Caucasian children at 0.15 mg per week. This difference is probably a result of the fact that the Caucasian children consumed more food and drink than did the Asian children. The Asian children had a significantly lower ($p < 0.05$) geometric mean blood lead concentration than the Caucasian group. There was no correlation between blood lead concentrations and lead intakes but it was found that children who washed their hands before eating meals or snacks tended to have the lowest blood lead concentrations. No correlation was found between blood lead concentrations and plasma concentrations of the 25-hydroxy metabolite of vitamin D, contrary to a previous report in the literature[34].

50.   Lastly a duplicate diet study of pre-school age children was carried out in Birmingham as part of a wider survey designed to estimate exposure to lead from various sources, such as air, dust, soil, food and water[37]. The mean dietary lead intake was estimated at 0.19 mg per week. This was equivalent to 0.02 mg per kg body weight per week, and 9 per cent of the children marginally exceeded the FAO/WHO Provisional Tolerable Weekly Intake for children of 0.025 mg per kg body weight per week for the period of study[38]. Data were also collected on the amount of canned food the children ate, but there was no correlation between this and blood lead levels.

51.   Children's lead intakes from the 3 studies where water lead levels were relatively low were 0.11 mg per week (control study), 0.15 mg per week (Harrow study) and 0.19 mg per week (Birmingham study). Although the children in Birmingham ate less on average than children in the other 2 surveys, their mean lead intake was higher. This may be explained by water lead concentrations being higher in Birmingham than in Harrow (no data were available on water lead levels in the control study).

52.   When duplicate diet studies are carried out participants collect duplicate portions of all the food and drink consumed and record the weights of the items in a diary over a period of one week. It was found that the diaries generally recorded a higher mean weight of both food and beverages than was measured by the laboratory prior to analysis. For the control study the relationship was found to be

Weight recorded in diary $= 1.01 + 0.9 \times$ observed weight.

Since the estimated lead intake depends on the weight of diet collected, duplicate diets tend to give an underestimate of intake, as noted in para. 46.

Interrelationships of Lead Levels in the Diet, Water and Blood

53.   As indicated in our previous report[2], the contribution made by inhalation of lead to blood lead is normally much less than that made by ingestion of food and water. The following comments therefore exclude the contribution from inhaled lead. The concentration of lead in whole blood is considered to be one of the best indicators in population groups of recent exposure to lead. Levels of blood lead were estimated by Piomelli *et al.*[39] in a remote population at the foothills of the Himalayas. The median blood lead concentration was 3.4 μg/100 ml. Work carried out in Dundee, where there is a relatively low exposure to lead from the environment, showed 1,665 mothers of newborn children to have a geometric mean blood lead concentration of 6 μg/100 ml[40]. Many data have been collected for populations exposed to specific sources of lead contamination, for example through occupational exposure or through living near to any lead producing or using industry. However, little information was available on the extent to which an individual's blood lead concentration would vary with time through exposure to environmental lead. To provide such data, the temporal stability of blood lead concentrations of 21 healthy adults exposed only

to environmental lead was assessed by analysis of 253 blood samples collected serially over periods of 7–11 months[41]. Blood lead concentrations in the serial samples from both men (mean: 12.2 μg/100 ml) and women (mean: 8.5 μg/100 ml) changed very little over the study period with a standard deviation of less than 0.5 μg/100 ml for the majority of individual mean concentrations. This stability confirms that a single value for blood lead concentration is a reliable indicator of lead exposure for environmental monitoring for adults.

54.   A previous duplicate diet study of infants carried out in Glasgow[42] had shown that there was a positive association between lead intake and blood lead concentration and that this relationship was described by the following equation:

$$PbB = 4.0 + 24 . \sqrt[3]{\text{week's intake (mg)}}$$

where PbB = Blood lead concentration in μg/100 ml.

The data from Ayr for infants supported this hypothesis. Indeed the cube root equation which best fitted the combined results for Ayr and Glasgow was:

$$PbB = 2.5 + 26 . \sqrt[3]{\text{week's intake (mg)}}$$

which is very similar over the relevant range to that for Glasgow alone.

55.   Dietary lead intake is related to the amount of lead in the water. The results for *adults* from Ayr were best described by the equation:

$$\text{Intake (mg/wk)} = 0.62 + 5.6 \, PbW(mg/l)$$

where PbW = Water lead concentration.

That is since the estimated mean intake was 2.5 mg per week, only about 0.6 mg per week of lead is derived from sources other than water, i.e. the diet (compared to an average UK value of 0.42 mg per week [para.22] from the diet). This is in an area with a plumbosolvent water supply and lead pipes. For the Birmingham *children*, where water lead levels were not unduly elevated, the regression equation was:

$$\text{Intake (mg/wk)} = 0.133 + 1.8 \, PbW(mg/l)$$

The mean lead intake was estimated as 0.19 mg per week so the contribution made by water lead to overall intake was lower, as would be expected.

56.   After 1981 the water supply in Ayr was treated to reduce its plumbosolvency. A follow-up study was carried out to see whether this action had had any effect on water lead and blood lead concentrations[43]. Before treatment of the water supply, the equation which best fitted the results was:

$$\text{Blood lead } (\mu g/100 \text{ ml}) = 4.7 + 27.8 . \sqrt[3]{PbW} \text{ kettle (mg/l)}$$

where PbW kettle = Kettle water lead concentration.

Combining the data with those obtained in the follow up study gives:

$$\text{Blood lead } (\mu g/100 \text{ ml}) = 5.6 + 26.2 . \sqrt[3]{PbW} \text{ kettle (mg/l)}$$

The curvilinear relationship implies that successive decreases in water lead concentrations will yield progressively larger decreases in blood lead concentrations. Thus even relatively low concentrations of lead in water would be expected to have a marked effect on the concentration of lead in blood. It follows that one strategy to obtain low blood lead concentrations would be to reduce the exposure to lead for drinking water as far as is practicable. Using data derived from the Ayr study[31] and the Glasgow study[42] where water lead concentrations of up to 1,400 μg/l were obtained, an increase of blood lead of 3.5–7 μg/100 ml has been estimated to result from an intake of 0.1 mg per day of lead from tap water.

57.   Current Department of Health advice is that where a person—particularly a child—has a blood lead concentration over 25 μg/100 ml, their environment should be investigated and steps taken to reduce exposure[44]. The EC Directive relating to the quality of water intended for human consumption[45], which came into effect in July 1985, sets a Maximum Admissible Concentration (MAC) for lead of 50 μg/l (in running water). However, there is an ambiguously worded qualifying comment. The UK interpretation of the lead parameter of the Directive[46] has been that for the MAC 50 μg/l must not be exceeded in routine samples taken after fully flushing, and taken from the water supply zone as a whole over a period of time. In addition,

   (i)   action should be taken in any supply zone where more than 2.5 per cent of samples exceed 100 μg/l[47], and

   (ii)   action should be taken in individual houses where a 30 minute stagnation sample exceeds 100 μg/l.

In the early 1980s water undertakers identified 130 zones serving nearly 5 million people where action was required in accordance with the criterion in (i). At the end of June 1989 remedial water treatment action had been completed successfully in 110 zones serving almost 4 million people; remedial action is underway in the remaining zones and is expected to be completed by the end of 1989.

The EC Directive has now been implemented in England and Wales through the Water Supply (Water Quality) Regulations 1989*. As a consequence of recent medical advice† the Regulations set more stringent requirements than the Directive. The lead standard in the Regulations is 50 μg/l in any sample and the ambiguous comments in the Directive relating to flushed samples and 100 μg/l have not been included in the Regulations. The Regulations place new responsibilities on water undertakers. Whenever there is a risk that the standard may be exceeded in a water supply zone they are required to treat the water except where:

   (i)   the treatment is unlikely to achieve a significant reduction in the concentration of lead; or

---

*The Water Supply (Water Quality) Regulations 1989. SI No 1147. HMSO, London.
†Water Policy Letter WP17/1989. Department of the Environment and Welsh Office.

(ii) the prescribed risk relates only to water supplied in an insignificant part of the zone; or

(iii) treatment is not reasonably practical.

The water undertaker also has a duty to remove its part of a lead pipe where the remainder of that lead pipe connecting to a drinking water tap under mains pressure has been removed and the owner of the premises has requested the undertaker in writing to remove its lead pipe. Guidance on the implementation of these responsibilities is set out in the document "Guidance on Safeguarding the Quality of Public Water Supplies"*. This states that a risk should be presumed in all water supply zones unless there is clear evidence to the contrary. Thus, treatment or further treatment is likely to be required in some water supply zones to reduce lead concentrations.

58.   From the results of the Glasgow and Ayr Duplicate Diet Studies[2, 31, 32, 48, 49] it is possible to estimate an upper limit to the concentration of lead in water that would enable a given standard for blood lead to be achieved. It is, however, necessary to take into account the variability of blood lead in the population arising both from differing exposure to other sources of environmental lead[48] and from the differences in individuals' intake, uptake and metabolism of lead[50, 51]. Results from the duplicate diet studies and unpublished data on blood lead levels for people exposed to about 50 μg/l of lead in water both indicate that for adults, average water lead concentrations should not exceed about 30 μg/l. For the critical group, bottle-fed infants (90 per cent of whose diet is water) average water lead concentrations should not exceed 10–15 μg/l. This figure is derived as follows:

The FAO/WHO Provisional Tolerable Weekly Intake (PTWI) for infants and children is 25 μg/kg body weight[38]. In the Ayr Duplicate Diet Study[31, 32], the average weight of the infants was 5.5 kg and the average diet weighed 6.25 kg. The concentration of lead in infant feed is 50–80 μg/kg[2, 4] and the proportion by weight of dried milk to water in infants' feed is about 1:8. Thus the PTWI implies that in the absence of other sources of lead, the concentration of lead in the water used to make up the feed must not be more than is given by the expression

$$\frac{25 \times 5.5}{6.25} = 1/9 \times 80 + 8/9 \times (\text{concentration of lead in water})$$

i.e. must not exceed 15 μg/l. Making allowance for exposure to lead from dust and in the air, the PTWI would probably just be met at an average water lead concentration of 10 μg/l[52].

Comparison of Dietary Intakes of Lead in the UK with those of Other Countries

59.   In the design of any dietary survey there are a great many variables, because it is unlikely that any 2 surveys will be carried out in exactly the same

*Guidance on Safeguarding the Quality of Public Water Supplies. HMSO. London (in press).

way. Therefore comparison of data obtained from different surveys may be misleading if the effects of differences in methods are not carefully assessed. Several factors can influence the estimate of metal intake, including the type of study (total diet or duplicate diet) (para.46), assumptions made about metal concentrations which are reported as being less than the LOD (para.19), the degree of preparation of foodstuffs, whether or not they are cooked prior to analysis (para.60), the accuracy of the analytical methods used, and the inclusion or exclusion of lead intake from drinking water.

60.   It has been shown that a very high proportion of lead can be removed from vegetables by careful preparation and cooking (para.41). This is especially true for root vegetables which are more likely to be contaminated with soil than leafy ones. It follows that countries where total diet samples are analysed as purchased may appear to have a higher lead intake than those countries where samples are prepared and cooked prior to analysis.

61.   In view of the difficulties in making valid comparisons it is prudent not to place too much emphasis on the actual figures reported but rather to consider whether any countries have markedly high or low intakes and whether both duplicate and total diet studies show the same pattern of results. Data reported in the literature on lead intakes of adults in the UK and other countries[53-64] are given in Table 9 of Appendix IV. Care is required in interpreting these data because of the differences in methodology used in the various countries. This is a good example of the need to report details of the quality assurance of experimental results.

62.   The data confirm that duplicate diet studies tend to give lower estimates of lead intake than do total diet studies. However both types of study rank the UK at the lower end of the range of estimated intakes. Whether or not this is merely a reflection of the fact that foods were prepared and cooked, and that low LODs were achieved with the analytical methods used is difficult to judge.

Contribution of Airborne Lead to Dietary Intake and Blood Lead

63.   In *Survey of Lead in Food: Second Supplementary Report*[2] the average proportion of dietary lead derived from atmospheric fallout was estimated to be between 13 and 16 per cent for adults. By means of balance studies the contribution of airborne lead to dietary intake was assessed in a study at Westminster Hospital, London. Stable lead and ^{210}Pb were determined in duplicate diets and faecal samples for handicapped children. Airborne lead contribution was then calculated by comparison with fallout data using the same method as that described in *Survey of Lead in Food: Second Supplementary Report*[2]. Making the same assumptions about fallout levels as in the study in that report, the average airborne contribution was 25–31 per cent. However, using data more appropriate for rural areas gave an estimate of 13–16 per cent. Thus the contribution lies between 13 and 31 per cent.

19

64. A study in northern Italy involved the changing of the isotope ratio of ^{206}Pb:^{207}Pb of the lead in petrol in a well-defined geographical area and observation of blood lead isotopic ratio levels. The Isotopic Lead Experiment[65] proved unfortunately, to have important weaknesses[66]. Nevertheless, results for a very small number of adult subjects (who may not have been representative[67]) indicated that in the conditions obtaining in Turin, where traffic levels were high and congestion acute, wind speeds were low and air lead concentrations were extremely high (averaging 4–6 μg/m^3). The proportion of lead in blood derived from petrol lead was about 25 per cent. In rural areas around Turin, the proportion was around 15 per cent. Although it was originally not entirely clear whether equilibrium in blood lead had been reached after the changeover in petrol lead, further analysis of the results using a dynamic model has confirmed these figures[68].

Uptake and Excretion of Lead by Humans

Uptake

65. Data submitted to the RCEP at the time the Royal Commission was studying lead in the environment showed that there was a large imbalance between estimates of faecal excretion of lead reported in the literature and the Working Party's estimates of lead intake, faecal excretion apparently being the larger quantity at about 0.2 mg per day. In order to investigate this discrepancy several studies have been carried out by AERE, Harwell. The first study investigated the uptake of lead by humans and how the rate of uptake is influenced by minerals and food[69]. Fasting subjects absorbed 40–50 per cent of ^{203}Pb taken as the chloride in distilled water. However, when the lead was ingested as part of a meal, uptake was only 7 per cent. Uptake of ^{203}Pb was also reduced by the addition of minerals. When taken with tea or coffee, uptake averaged 14 per cent and with beer 19 per cent. Thus lead is absorbed from beverages at approximately twice the rate at which it is taken up from food.

Excretion

66. In a subsequent study 10 volunteers collected their total output of urine and faeces for one week. For those volunteers who were abstemious, mean daily excretion of lead was 0.07 ± 0.01 mg. For those volunteers who drank alcoholic beverages, mean daily excretion of lead was 0.16 ± 0.04 mg. The difference between the means is statistically significant. Lead intake from the British Total Diet Study in recent years is estimated at between 0.02 and 0.07 mg per day, with another 0.01 mg per day coming from air and water. Thus for abstemious subjects the agreement is reasonable. However, for subjects who drank alcoholic beverages there was clearly an imbalance. Since the British Total Diet Study does not include alcoholic beverages and since it is known that both beer and wine may be contaminated with lead[21], this may well be the source of lead intake for 'drinking' subjects.

67.   A further study followed up these results by measuring ^{210}Pb in the urine and faeces of the volunteers from the previous study (para.66). Total ^{210}Pb excretion by the abstemious subjects was 0.05 ± 0.01 Bq per day, and for those drinking alcoholic beverages it was 0.06 ± 0.02 Bq per day. This difference is not significant. The average ^{210}Pb content[2] of total diet samples is 0.05 Bq per day. Thus the total excretion of lead in *all* the volunteers was very similar to the estimated total dietary intake of ^{210}Pb. This contrasts with the situation for stable lead and suggests that the source from which drinkers of alcoholic beverages take their additional lead has a low ratio of Bq ^{210}Pb per g to stable lead (i.e. it is derived from mineral rather than atmospheric sources).

68.   Literature data[69, 70] suggest that a link exists between alcohol consumption and blood lead levels, with extreme consumers of alcoholic beverages having elevated levels of lead in their blood. A survey of wine and beer[21] has shown that lead contamination of alcoholic beverages can occur, and it has also been shown that lead from beer is more readily absorbed than lead from food. Mean consumption of beer for adults in the UK is estimated to be 0.7 l per day and extreme consumption 2.1 l per day. Assuming low level lead contamination of beer of 0.05 mg/l, then extreme intake of lead from beer would be 0.11 mg per day of which 20 per cent (0.02 mg per day) will be absorbed. Lead intake from the rest of the diet currently has an estimated maximum value of 0.06 mg per day (para.20) of which less than 10 per cent is absorbed (i.e. a maximum of 0.006 mg per day). Thus heavy drinking of beer may cause a substantial increase in lead intake and elevated levels of lead in the blood.

69.   A study was carried out by AERE, Harwell to assess the average increase in blood lead level resulting from daily drinking of beer supplemented with lead to the highest levels found in the 1982 survey[21]. Blood samples were taken at regular intervals from 9 volunteers who consumed beer containing 0.43 mg/l of lead. A control group consumed beer containing 0.005 mg/l of lead. Figure 1 shows the mean increases in blood lead with time for subjects and controls. The mean increase was 2.8 ± 0.7 μg/100 ml for every increase of 0.1 mg in daily intake of lead. This is equivalent to an equilibrium value of 4.2–5.6 μg/100 ml for every increase of 0.1 mg in daily intake of lead. These data are very similar to those determined for drinking water with elevated levels of lead (para.56).

Uptake of Lead by Plants from the Soil and from Atmospheric Deposition

Uptake from the soil

70.   Information on crop/soil relationships for lead is usually expressed in terms of concentration factors (CF), which may be defined as:

$$CF = \frac{\text{amount of lead in plant tissue (dry weight or wet weight)}}{\text{amount of lead in soil (dry weight or wet weight)}}$$

When measuring CFs it is important to distinguish between lead which has been taken up by roots and that derived from soil contamination or aerial deposition.

Fig 1. **Increase in Blood Lead Concentration following consumption of beer supplemented with Lead.**

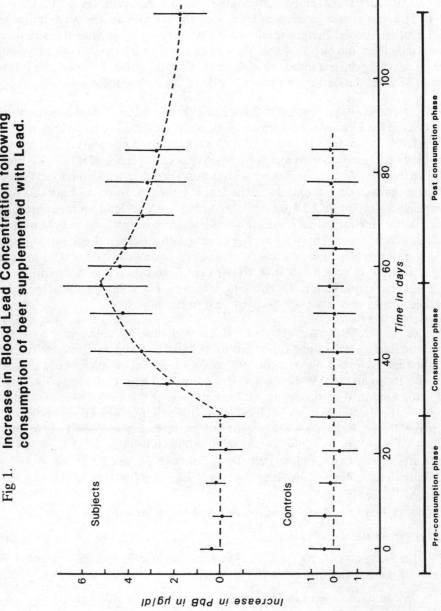

One means by which this can be achieved involves the addition of a ^{210}Pb tracer to the soil. Experiments carried out at AERE, Harwell for cabbage, carrot and spinach grown on soils with a pH of 7 or above suggest that the CFs in the edible parts are very small, $0.3-2 \times 10^{-3}$. Most background soils in the UK contain lead in the range 10–150 mg/kg[5] and the median value is 40 mg/kg. Assuming a CF of 2×10^{-3}, then the corresponding amount of lead in the plant is 0.08 mg/kg dry weight or 0.008 mg/kg fresh weight. This suggests that other routes of contamination, soil splash and aerial deposition, are more important for plants on normal agricultural soils. CF values for plants grown on acidic sandy soils ranged from 6×10^{-3} to 33×10^{-3}, an order of magnitude higher than those obtained for neutral soils.

71. The distribution of lead between outer and inner leaves of leafy crops was also investigated. More lead was found in the outer leaves with an average outer/inner ratio for cabbage of about 5 and for spinach of about 2.5. This could have been due to an age effect reflecting the length of exposure to airborne lead. Alternatively, this distribution may simply reflect greater transpiration from the outer leaves than from the tightly formed heart of cabbage. Spinach, on the other hand, has a more open heart and thus the rate of transpiration from these tissues will be closer to that for the outer leaves. These findings emphasise the prudence of trimming vegetables closely before cooking and eating.

Lead uptake in crops from soils treated with sewage sludge

72. Soils treated with sewage sludge can accumulate lead which may be taken up by crops grown on these soils. An EC Directive[71] limits to 300 mg/kg maximum lead concentrations in soils treated with sewage sludge and revised national guidelines are shortly to be issued in the UK to implement this Directive*. In order to aid definition of UK guidelines for soil metal limits on sludge-treated land, a 5–year crop trial sponsored by the Department of the Environment was carried out[72] between autumn 1978 and spring 1983. The main objectives were to determine the uptake of metals by crops grown under different conditions and to observe any effect of time on uptake. Four types of digested sludge and 6 rates of application were used, the latter to give the required range of metal levels in the soil. Six crops were studied to reflect the most important constituents of the human diet (wheat, potatoes, cabbage), metal-sensitive indicator crops (red beet and lettuce) and rye-grass. The highest soil lead concentrations in the trial were 330 mg/kg. Lead concentrations in most of the crops in the trial did not significantly rise above 0.5—2.0 mg/kg dry weight. The exception to this was lettuce where mean concentrations were about 5 mg/kg, corresponding to a fresh weight concentration of 0.2mg/kg. In general, no significant rise in lead concentrations occurred in crops above background levels and lead was therefore relatively unavailable to crops from the types of soil

---

*Now issued as "Code of Practice for Agricultural Use of Sewage Sludge." (1989). HMSO. London. ISBN 0 11 752256 2.

investigated. Overall there was no trend in lead availability from soil to crops over 5 years. However, there was little evidence that sludge metals were lost from the cultivated layer except where sludge was originally incorporated below this depth.

Lead fall-out from the atmosphere

73. Foliar uptake and redistribution of lead were measured by AERE, Harwell in 3 species of plant (carrot, french bean and radish). A solution containing ^{210}Pb (7.4×10^5 Bq ^{210}Pb/mg Pb) was applied directly to the leaf surfaces, stringent precautions being taken to avoid contamination of the growing medium. The proportion of applied lead transported to the root tissue was small for both radishes (less than 0.3 per cent) and carrots (1 per cent). In the case of beans no activity was detected in the seed or pod tissue. The contribution of foliar absorbed lead to the total lead burden may be estimated. For example, calculations for one carrot estimate the total fall-out as 0.12 mg. This assumes a growing period of 15 weeks, and a fall-out rate of 20 mg/m^2 per year^2 for a plant of 0.02 m^2 leaf area. If 0.1 per cent of this fall-out is transported to the root organ with a dry weight of 3 g (30 g fresh weight), the lead burden would be equivalent to 0.024 mg/kg dry weight or 40 per cent of the total carrot root lead burden. In areas with higher lead fallout, the relative contribution of foliar absorbed lead is likely to be similar because of greater associated soil burden and uptake.

74. A second study carried out at the University of Nottingham indicated, however, that more than 97 per cent of the lead content of carrot roots was derived from the atmosphere. In these experiments a different technique was used to estimate the contribution made by air lead. Lead uptake from the soil by carrot roots and leaves was determined by growing the carrots in soil containing 2.2×10^5 Bq ^{210}Pb per kg soil. Assuming uptake of stable lead to be the same as that of ^{210}Pb, the atmospheric input of stable lead was determined by difference. Ninety nine per cent of the foliar leaf lead was derived from the atmosphere and 8 per cent of this deposited lead was transported down to the root.

75. The reasons for the discrepancy between the 2 estimates are not clear and further work would be necessary to resolve this point. At present the best estimate of the contribution made by air lead to lead concentration in root crops is 40–100 per cent.

Soil Contamination

76. Problems owing to soil contamination have been recognised for some time, and plant material is therefore thoroughly washed and prepared prior to analysis. However, even when these precautions have been taken, plant samples containing unusually high lead concentrations may still be found, implying that

24

soil contamination does not necessarily have to be visible to make a significant contribution to the lead content. Some of the vegetables analysed as part of the extensive survey carried out at Shipham in Somerset[18] contained very high apparent concentrations of lead (more than 200 mg/kg on a dry weight basis). Assuming the CF for the mineralised and calcareous Shipham soils to be 1×10^{-2} and taking the mean concentration of lead in the soils in which the vegetables were grown to be 4,000 mg/kg, then the vegetables should not have contained more than 40 mg/kg of lead.

77.   A number of vegetable samples were acid digested and soil contamination assessed by whether or not there were traces of insoluble solid material in the digests. There was a clear difference between the lead concentrations of those vegetables containing a residue and those which did not. Most of the samples containing a residue were leafy vegetables and it may be that leafy vegetables are more prone to soil contamination than are other vegetables. The levels of contamination observed in these samples (mean lead concentration: 285 mg/kg) could be accounted for by the presence in 10 g of fresh sample of only 36 mg of soil containing 4,000 mg/kg of lead.

Speciation of Lead in Food

78.   Whilst there is substantial literature on the occurrence of lead in food there is very little information about the chemical form or forms in which the metal is present. This no doubt reflects the considerable analytical difficulties in reliably measuring very low levels of the individual lead compounds in a complex biological matrix.

79.   Lead has 3 oxidation states $(0, +2$ and $+4)$ and in addition to numerous inorganic complexes, organometallic species are also known. The occurrence of the free metal (oxidation state 0) in foods is likely to be restricted to the presence of lead shot in game birds[73]. In aqueous solution the $+2$ oxidation state predominates except under strongly oxidising conditions[74]. It is therefore probable that the soluble lead species in foodstuffs will largely comprise hydrated plumbous species complexed with soluble ligands from the food matrix, such as amino and carboxylate groups and chloride ions. In addition to these soluble species a proportion of the cationic lead in foodstuffs will probably be associated with the insoluble fraction. This is likely to arise from formation of insoluble lead salts such as oxides and carbonates and also complexation of lead with water-insoluble ligands, including those in denatured and structural proteins.

80.   The occurrence of organometallic species in the environment is considerably less than that of inorganic lead[75, 76, 77]. As a result, it is probable that these compounds do not contaminate foodstuffs to a significant extent. Tetraethyl lead is widely used as an anti-knock agent in petrol. However, during combustion it is largely converted to inorganic oxides, carbonates and halides[78]. It has

been suggested that organometallic compounds may be formed in sediments by biological methylation of inorganic lead[79]. This phenomenon has not been observed in other studies[80, 81] and there is no reason to suppose that the putative formation of these compounds in soil sediments results in contamination of the food chain. In conclusion there is little known about the exact chemical speciation of lead in foods. What evidence there is suggests that inorganic lead is the major form and that this exists in a range of soluble and insoluble cationic species.

CONCLUSIONS

81.   It is estimated from the British Total Diet Study that the dietary intake of lead from food and beverages, excluding tap water, in Great Britain is currently between 0.02 and 0.06 mg per day. There has been a gradual decrease in lead concentrations in those food groups containing predominantly canned food (canned vegetables and fruit products) since 1982. **(Paras.15–22.)**

82.   Lead concentrations in most individual foods were generally low. All but a few samples contained lead at levels below those prescribed in *The Lead in Food Regulations, 1979*. Elevated lead levels have been found in some poured wines and steps are now being taken to reduce these levels. **(Paras.23–40.)**

83.   Four dietary studies to estimate individual lead intakes have been carried out since the last report[2]. In areas where water lead concentrations were high, elevated dietary lead intakes were found for both adults and children. A control dietary study showed that levels of lead found in diets of both adults and children were comparable to those found from the British Total Diet Study. **(Paras.43–52.)**

84.   Three studies have been carried out on the uptake of lead since the last report. Absorption of lead from beverages was found to be approximately twice that from food, for adults. **(Paras.65–69.)**

85.   Lead is not taken up readily by plants from soils; concentration factors for acidic sandy soils although still very small are, however, an order of magnitude higher than those obtained for other soils. Lead was relatively unavailable to crops from soils treated with sewage sludge at soil lead concentrations up to 330 mg/kg. **(Paras.70–72.)**

86.   The contribution made by lead in air to dietary lead remains uncertain. **(Para.63.)** Similarly, the contribution of air lead to the lead content of root crops has yielded widely varying proportions, of between 40 and 97 per cent. **(Paras.73–75.)** It has not been considered necessary to pursue the work on lead in crops further due to the currently decreasing importance of air as a source of lead exposure. Little is known about the chemical forms of lead in food and further work will be carried out on this important topic. **(Paras.78–80.)**

ACKNOWLEDGEMENTS

The analytical studies and surveys reported here were carried out at laboratories of the following establishments and organisations:

AFRC Institute of Food Research, Norwich Laboratory

Agricultural Development Advisory Service, Ministry of Agriculture, Fisheries and Food

Atomic Energy Research Establishment, Harwell

Avon County Scientific Services, Bristol

Campden Food Preservation Research Association, Gloucs.

Charing Cross and Westminster Medical School, London

Directorate of Fisheries Research, Ministry of Agriculture, Fisheries and Food

Food Science Laboratory, Ministry of Agriculture, Fisheries and Food

Laboratory of the Government Chemist, Department of Trade and Industry

Long Ashton Research Station, Bristol

Research Services Ltd., London

University of Nottingham

University of Southampton

REFERENCES

1. World Health Organisation (1972) Evaluation of Certain Food Additives. Sixteenth Report of the Joint FAO/WHO Expert Committee on Food Additives. *FAO Nutrition Meetings Report Series No. 51*, publ. FAO and WHO.

2. Ministry of Agriculture, Fisheries and Food (1982) Survey of Lead in Food: Second Supplementary Report. *Food Surveillance Paper No.* **10,** publ. HMSO.

3. Ministry of Agriculture, Fisheries and Food (1972) Survey of Lead in Food. Working Party on the Monitoring of Foodstuffs for Heavy Metals. Second Report, publ. HMSO.

4. Ministry of Agriculture, Fisheries and Food (1975) Survey of Lead in Food. First Supplementary Report. Working Party on the Monitoring of Foodstuffs for Heavy Metals. Fifth Report, publ. HMSO.

5. Royal Commission on Environmental Pollution (1983) Lead in the Environment. Ninth Report, publ. HMSO.

6. Ministry of Agriculture, Fisheries and Food (1983) Report on the Review of Metals in Canned Foods. **FAC/REP/38,** publ. HMSO.

7. Ministry of Agriculture, Fisheries and Food (1988) The British diet: finding the facts. *Food Surveillance Paper No.* **23,** publ. HMSO.

8. *The Lead in Food Regulations 1961, S.I. [1961] No. 1931*, publ. HMSO.

9. *The Lead in Food Regulations 1979, S.I. [1979] No. 1254*, publ. HMSO.

10. *The Lead in Food (Amendment) Regulations 1985, S.I. [1985] No. 912*, publ. HMSO.

11. *The Natural Mineral Waters Regulations 1985, S.I. [1985] No. 71*, publ. HMSO.

12. *The Motor Fuel (Lead Content of Petrol) (Amendment) Regulations 1985, S.I. [1985] No. 1728*, publ. HMSO.

13. Evans, W.H., Read, J.I. and Lucas, B.E. (1978) Evaluation of a method for the determination of total cadmium, lead and nickel in foodstuffs using measurement by flame atomic absorption spectrophotometry. *Analyst* **103,** 580–594.

14. Ministry of Agriculture, Fisheries and Food (1984) Steering Group on Food Surveillance Progress Report 1984. *Food Surveillance Paper No.* **14,** publ. HMSO.

15. Holden, A.V. and Topping, G. (1981) Report on further intercalibration analysis in ICES pollution monitoring and baseline studies, Co-operative Research Report No. **108.** Copenhagen: International Council for the Exploration of the Sea.

16. Knowles, M.E., Burrell, J.A. and McWeeny, D.J. (1983) Joint FAO/WHO food contamination monitoring programme, analytical quality assurance. 11. Geneva, WHO.

17. Sherlock, J.C., Evans, W.H., Hislop, J., Kay, J., Law, R., McWeeny, D.J., Smart, G.A., Topping, G. and Wood, R. (1985) Analysis—accuracy and precision? *Chemistry in Britain* **21**, 1019–1021.

18. Sherlock, J.C., Smart, G.A., Walters, B., Evans, W.H., McWeeny, D.J. and Cassidy, W. (1983) Dietary surveys on a population at Shipham, Somerset, United Kingdom. *Sci. Tot. Environ.* **29**, 121–142.

19. Chattopadhyay, A. (1974) Ph.D. Thesis, University of Toronto, Canada.

20. Crosby, W. (1977) Lead contaminated health food. Association with lead poisoning and leukemia. *J. Amer. Med. Assoc.* **237**, 2627–2629.

21. Sherlock, J.C., Pickford, C.J. and White, G.F. (1986) Lead in alcoholic beverages. *Food Additives and Contaminants* **3**, 347–354.

22. Simpson, A.C. (1982) Alternatives to cork and glass. Paper presented at the Institute of Masters of Wines' International Symposium on Viticulture, Unification and the Treatment and Handling of Wine, Oxford.

23. Wai, C.M., Knowles, C.R. and Keely, J.F. (1979) Lead caps on wine bottles and their potential problems. *Bull. Environ. Contam. Toxicol.* **21**, 4–6.

24. Smart, G.A., Pickford, C.J. and Sherlock, J.C. (1989) Lead in alcoholic beverages: a second survey. *Food Additives and Contaminants.* (In press).

25. Ministry of Agriculture, Fisheries and Food (1987) "Wipe before you pour says Donald Thompson" *Food Facts Press Release* **FF 18/87.**

26. Warwick, M.E. (1980) Laboratory studies of the corrosion of side seams in soldered tinplate containers. 2nd International Tinplate Conference, London, October 1980.

27. Capar, S.G. (1978) Changes of lead concentrations of foods stored in their open cans. *J. Food Safety* **1**, 241–245.

28. Smart, G.A. and Sherlock, J.C. (1989) Survey of canned food storage in the home. *Food Additives and Contaminants* **6**, 125–132.

29. Woman's Own. Getting the best out of cans. 17 October 1985.

30. Smart, G.A., Warrington, M. and Evans, W.H. (1981) The contribution of lead in water to dietary lead intakes. *J. Sci. Fd. Agric.* **32**, 129–133.

31. Sherlock, J.C., Smart, G.A., Forbes, G.I., Moore, M.R., Patterson, W.J., Richards, W.N. and Wilson, T.S. (1982) Assessment of lead intakes and dose-response for a population in Ayr exposed to a plumbosolvent water supply. *Human Toxicol.* **1**, 115–122.

32.   Sherlock, J.C., Ashby, D., Delves, H.T., Forbes, G.I., Moore, M.R., Patterson, W.J., Pocock, S.J., Quinn, M.J., Richards, W.N. and Wilson, T.S. (1984) Reduction in exposure to lead from drinking water and its effect on blood lead concentrations. *Human Toxicol.* **3**, 383–392.

33.   Kim, W., Hertz, W., Judd, J., Marshall, M., Kelsay, H. and Prather, E. (1984) Effect of making duplicate food collections on nutrient intakes calculated from diet records. *Amer. J. Clin. Nutrition* **40**, 1333–1337.

34.   Box, V., Cherry, N., Waldron, H.A., Dattari, J., Griffiths, L.D. and Hill, F.G. (1981) Plasma vitamin D and blood lead concentrations in Asian children. *Lancet* August 15, 373.

35.   Department of the Environment, Steering Committee on Environmental Lead in Birmingham (1982) Blood lead concentrations in pre-school children in Birmingham. *Pollution Report No.* **15,** Department of Environment, London.

36.   Sherlock, J.C., Barltrop, D., Evans, W.H., Quinn, M.J., Smart, G.A. and Strehlow, C. (1985) Blood lead concentrations and lead intake in children of different ethnic origin. *Human Toxicol.* **4,** 513–519.

37.   Smart, G.A., Sherlock, J.C. and Norman, J.A. (1988) Dietary intake of lead and other metals. A study of young children from an urban population in the UK. *Food Additives and Contaminants* **5,** 85–93.

38.   WHO, (1987) Evaluation of Certain Food Additives and Contaminants. Thirtieth Report of the Joint FAO/WHO Expert Committee on Food Additives. WHO Technical Report Series No. **751.** World Health Organisation, Geneva.

39.   Piomelli, S., Corash, L., Corash, M.B., Seaman, C., Mushak, P., Glover, B. and Padgett, R. (1980) Blood lead concentrations in a remote Himalayan population. *Science* **210,** 1135–1137.

40.   Zarembski, P.M., Griffiths, P.D., Walker, J. and Goodall, H.B. (1983) Lead in neonates and mothers. *Clin. Chim. Acta* **134,** 35–49.

41.   Delves, H.T., Sherlock, J.C. and Quinn, M.J. (1984) Temporal stability of blood lead concentrations in adults exposed only to environmental lead. *Human Toxicol.* **3,** 279–288.

42.   Department of the Environment (1982) The Glasgow duplicate diet study (1979/1980). *Pollution Report No.* **11,** Department of the Environment, London.

43.   Sherlock, J.C., Ashby, D., Delves, H.T., Forbes, G.I., Moore, M.R., Patterson, W.J., Pocock, S.J., Quinn, M.J., Richards, W.N. and Wilson, T.S. (1984) Reduction in exposure to lead from drinking water and its effect on blood lead concentrations. *Human Toxicol.* **3,** 383–392.

44. Joint Circular Department of the Environment/Welsh Office (1982) Lead in the Environment. Circular **22/82** (DoE)/Circular **31/82** (Welsh Office), publ. HMSO.

45. European Community (1980) Directive relating to the Quality of Water intended for Human Consumption (80/778/EEC). *Official Journal of the European Communities*, **L229/11–29**.

46. Department of Environment. Standing Technical Advisory Committee on Water Quality. Fourth Biennial Report. February 1981–March 1983. HMSO. London. 1984.

47. Department of the Environment. Lead in Potable Water. Report of the Expert Advisory Group on Identification and Monitoring (EAGIM). Technical Note No. 1. (Unpublished).

48. Lacey, R.S., Moore, M.R. and Richards, W.N. (1985) Lead in water, infant diet and blood: the Glasgow Duplicate Diet Study. *Sci. Tot. Environ.* **41,** 253–257.

49. Sherlock, J.C. and Quinn, M.J. (1986) Relationship between blood lead concentrations and dietary lead intake in infants: The Glasgow Duplicate Diet Study 1979–80. *Food Additives and Contaminants* **3,** 167–176.

50. James, H.M., Hilburn, M.E. and Blair, J.A. (1985) Effect of meals and mealtimes on uptake of lead from the gastro-intestinal tract in humans. *Human Toxicol.* **4,** 401–407.

51. Heard, M.J., Chamberlain, A.C. and Sherlock, J.C. (1983) Uptake of lead by humans and effect of minerals and food. *Sci. Tot. Environ.* **30,** 245–253.

52. Quinn, M.J. and Sherlock, J.C. (1990). The correspondence between UK 'action levels' for lead in blood and water. *Food Additives and Contaminants*. (To be published).

53. Boudene, C., Arsac, F. and Despaux, N. (1978) Research into food chain contamination by lead. Journal du Science. Evaluation du Research. *Research Environment* **14,** 339–344, Ministry of the Environment, Paris.

54. Kampe, W. (1983) Lead and cadmium in food—a current hazard? *Verbaucherdienst* **28,** 75–78.

55. Ministry of Health and Environmental Protection (1980) 'Man and Nutrition' Surveillance Programme. The Hague.

56. Jathar, V.S., Pendharkar, P., Raut, S. and Panday, V. (1981) Intake of lead through food in India. *J. Fd. Sci. Technol.* **18,** 240–242.

57. Lanzola, E. and Allergini, M. (1983) Ingestion of lead in the total diet—a study in an area of Northern Italy. *Riv. Soc. It. Sci. A.* **12,** 203–209.

58. Buchet, J.P., Lauwerys, R., Vandervoorde, A. and Pycke, J.M. (1983) Oral daily intake of cadmium, lead, manganese, copper, chromium, mercury,

calcium, zinc and arsenic in Belgium: a duplicate meal study. *Fd. Chem. Toxicol*. **21**, 19–24.

59.   Pfannhauser, W. and Woidich, H. (1980) Source and distribution of heavy metals in food. *Toxicol. Environ. Chem. Rev*. **3**, 131–144.

60.   Orlando, P. (1977) Research on the levels of microelements in ready to eat foods and total diet. Note V Lead. *G. Ig. Med. Prev*. **18**, 85–93.

61.   Fouassin, A. and Fondu, M. (1980) Estimation of the average intake of lead and cadmium from the Belgian diet. *Arch. Belg. Med. Soc. Hyg. Med. Trav. Med. Leg*. **38**, 453–467.

62.   Miljoministeriet (1980) Cadmiumforurening En redegorelse om anvendelse, forekomst og skadevirkninger af cadmium i Danmark. Copenhagen.

63.   Koivistoinen, K. (1980) Mineral element composition of Finnish foods. *Acta Agriculturae Scandinavica* **22**, 171.

64.   Gartrell, M.J., Crown, J.C. and Gunderson, E.L. (1986) Pesticides, selected elements and other chemicals in adult total diet samples. October 1980–March 1982. *J. Assoc. Off. Anal. Chem*. **69**, 146–161.

65.   Facchetti, S., Geiss, F., Gaglione, P., Colombo, A., Garibaldi, G., Spallanzani, G. and Gilli, G. (1982) Isotopic Lead Experiment—Status Report. EUR 8352 EN. Commission of the European Communities, Joint Research Centre, Ispra Establishment.

66.   Elwood, P.C. (1983) The Lead Debate. Proc. Conf. Institution of Environmental Health Officers (Brighton) 1983. London: IEHO.

67.   Elwood, P.C. (1983) Turin Isotopic Lead Experiment. *Lancet*, 16 April, 869.

68.   Fantechi, R. and Colombo, A. (1984) Isotopic Lead Experiment. A Dynamic Analysis of the Isotopic Lead Experiment Results. EUR 8760 EN. *European Applied Research Reports* **2**, 351–386.

69.   Shaper, A.G., Pocock, S.J., Walker, M., Wale, C.J., Clayton, B., Delves, H.T. and Hinks, L. (1982) Effects of alcohol and smoking on blood lead in middle-aged British men. *Brit. Med. Journal* **284**, 289–302.

70.   Grasmick, C., Huel, G., Moreau, T. and Sarmini, H. (1985) The combined effect of tobacco and alcohol consumption on the levels of lead and cadmium in blood. *Sci. Tot. Environ*. **41**, 207–217.

71.   European Community (1986) Directive on the protection of the environment, and in particular of soil, when sewage sludge is used in agriculture. (86/278/EEC). *Official Journal of the European Communities* **L181**, 6–12.

72.   Carlton-Smith, C.H., Stark, J.H., Thomas, B.A. and Post, R.D. (1987) Effects of Metals in Sludge on Crops. Final Report to the Department of the Environment. *Technical Report No. TR* 251. Water Research Council.

73. Moreth, F. and Hecht, H. (1981) Lead from residues of shot in game. *Fleischwirtschaft* **61**, 1315–1325.

74. Carr, D.S. (1981) Lead Compounds. Encyclopedia of Chemical Technology, Eds. I. Kirk and D.F. Othmer. Publ. John Wiley and Sons Ltd, pp. 160–180.

75. Alloway, B.J. and Tills, A.R. (1983) Speciation of lead in soil solution from very polluted soils. *Environ. Technol. Lett.* **4**, 529–534.

76. De Jonghe, W.R.A. (1981) Identification and determination of individual tetraalkyl lead species in air. *Environ. Sci. Technol.* **15**, 1217–1222.

77. Harrison, R.M. and Biggins, P.D.E. (1980) Chemical speciation of lead compounds in street dusts. *Environ. Sci. Technol.* **14**, 336–339.

78. Ter Haar, G.L. and Bayard, M.A. (1971) The composition of airborne lead particles. *Nature* **232**, 553–554.

79. Wong, P.T.S., Chan, Y.K. and Luxon, P.L. (1975) Methylation of lead in the environment. *Nature* **253**, 263–264.

80. Reisinger, K., Stoeppler, M. and Nurnberg, H.W. (1981) Evidence for the absence of biological methylation of lead in the environment. *Nature* **291**, 228–230.

81. Reisinger, K., Stoeppler, M. and Nurnberg, H.W. (1981) On the biological methylation of lead, mercury, methylmercury and arsenic in the environment. In: Heavy Metals in the Environment, CEP Consultants Ltd, pp. 649–652.

82. Medical Research Council (1988) The neuropsychological effects of lead in children. A review of the research 1984–88. London.

83. Baxter, M.J., Burrell, J.A., Crews, H.M., Massey, R.C. and McWeeny, D.J. (1989) A procedure for the determination of lead in green vegetables at concentrations down to 1μg/kg. *Food Additives and Contaminants* **6**, 341–349.

84. Department of the Environment (1988) *Digest of Environmental Protection and Water Statistics* No. **10**, publ. HMSO.

APPENDIX I CONSIDERATION OF THE REPORT BY THE COMMITTEE ON TOXICITY OF CHEMICALS IN FOOD, CONSUMER PRODUCTS AND THE ENVIRONMENT

1.   We have been requested by the Steering Group on Food Surveillance to consider the *Progress Report on Lead in Food, Third Supplementary Report*, compiled by the Working Party on Inorganic Contaminants in Food and to assess hazards to health that could arise from lead in foods in the context of present information on the toxicity of lead. We reported previously on lead in food in 1975 and 1982[2, 4].

2.   We note that it has been the policy of successive governments since 1974 to contain and reduce lead exposure wherever practicable. As part of this policy the Ministry of Agriculture, Fisheries and Food (MAFF) has taken action to reduce the degree of lead contamination of foods and beverages that can occur during their manufacture, processing and storage, and we *welcome* the detailed and extensive dietary monitoring surveys which have been carried out since 1974. In general, we find the gradually declining levels of lead in food reassuring. However, in view of the uncertainties surrounding the effects of low level lead exposure in children (see paragraphs 7 and 8), we believe it is prudent to continue to make every effort to reduce lead exposure from all sources, including food.

Dietary Lead Intake: Adults

3.   We *welcome* the drop in the average UK dietary lead intake as estimated by the Working Party compared with the estimate given in the Second Supplementary Report[2], although we note that these values are not directly comparable due to methodological changes in the preparation and analysis of diets which took place in 1981. In most foodstuffs, lead concentrations were below the analytical limits of detection and we *welcome* this. Exceptions were noted for offal, which is known to accumulate lead, some canned foods, some ethnic foodstuffs and bonemeal products.

4.   The dietary intake of lead in the UK is well below the amount that could cause classical manifestations of lead poisoning, the first signs of which are lethargy, anaemia and gastro-intestinal symptoms. This was reviewed in our report on the Second Supplementary Report. The current average dietary lead intake is also below the FAO/WHO Provisional Tolerable Weekly Intake (PTWI) for adults[1], which was based on the earliest symptoms of classical lead poisoning. There is a reasonable margin of safety between the average lead intake and the PTWI, which itself incorporates a large safety factor, for those consumers of large amounts of food or of unusual food items.

5.   Acute lead poisoning incidents can still occur on the premises of small works, such as battery breakers, where occupational hygiene may be inadequate. For the general population it is the long term accumulation of lead from a multiplicity of sources which is important. We note that for most adults in the UK food is the most important source of lead intake, while other sources such as air and drinking water make only a small contribution. Most adults have lead intakes from all non-occupational sources which are below the PTWI and do not, in our opinion, pose any risk of classical lead poisoning. Because of the importance of the diet as a source of lead intake we *recommend* that monitoring of lead in food as consumed should continue. We note that other sources such as drinking water and alcoholic beverages can raise lead intake to undesirable levels, and we are particularly concerned that the exposure of pregnant women to lead be kept to a minimum because of possible effects on the developing foetus.

Dietary Lead Intake: Infants and Children

6.   The results of the Duplicate Diet Studies carried out on pre-school children in different parts of the UK do not indicate that the estimated dietary lead intakes pose any risk of classical lead poisoning.

7.   In recent years concern about the toxicity of lead has centred on subtle neuropsychological effects of low level lead exposure in children. In 1988 the Medical Research Council's Lead Advisory Group (LAG) reported on the findings of recent epidemiological studies[82]. In its first report in 1983, LAG suggested that 'any effects of lead at the exposure levels seen in the UK are very small and cannot be detected with any certainty'. This conclusion is now considered to be 'still largely applicable but the evidence of an association between body lead burden and IQ is now stronger'.

8.   We note that the limitations of epidemiological studies in drawing causal inferences are such that it is not possible to conclude that exposure to current urban levels of lead is definitely harmful. However, since these studies do not provide reassurance as to the lack of harmful effects of lead we *recommend* that lead exposure be further reduced wherever practicable.

9.   The FAO/WHO Expert Committee considered the special factors which make children especially susceptible to the toxic effects of lead[38]. They established a PTWI for children from an intake level which did not produce an accumulation of lead in bottle fed infants. We *welcome* the results of the Working Party's Duplicate Diet Studies which indicate that the dietary lead intake for most children in the UK is below the PTWI. We note, however, that only a negligible margin exists between the current average dietary lead intake and the PTWI, and we are concerned that some children, especially bottle fed infants exposed to high drinking water lead concentrations are consuming

undesirable amounts of lead. For these groups drinking water is the predominant source of lead and we *recommend* that strenuous efforts continue to be made to reduce water lead levels in plumbosolvent parts of the UK.

Canned Food

10. We *welcome* the recent decline in the lead content of canned food, and consider that the removal of lead solder from cans manufactured in the UK has made an important contribution to reducing lead exposure.

11. We note that some imported cans still contain lead solder, and the degree of lead contamination of food can increase markedly when stored in opened cans of this type. In view of the Working Party's survey which found that 20 per cent of households store food in opened cans[28], we *recommend* that more effective publicity be given to consumers to discourage this practice.

Alcoholic Beverages

12. We have already expressed our concern that even modest consumption of wine from a lead capped bottle can substantially increase the normal dietary lead intake[25]. We note that lead consumed in beverages is more available for uptake than lead in food, and therefore reinforce our earlier *recommendation* that lead be substituted by an acceptable alternative for wine bottle seals.

13. We have been informed that representatives of the national and international wine trade are investigating ways of reducing the risks of lead contamination and we *welcome* this. As lead capped wine bottles will remain in use for some time, we *recommend* that MAFF continue to monitor lead levels and to advise consumers on a regular basis.

14. We *welcome* the recent remedial action taken by the UK Brewers Society which has reduced lead contamination of draught beers. We note that lead concentrations in bottled and canned beers are now generally low, but reaffirm our earlier *recommendation* that monitoring of all beers should continue.

Sources of Lead in Food

15. We are aware of public concern about the safety of lead intakes from vegetables grown in urban gardens, and have been informed that research into this matter is nearing completion. We note that the translocation of lead from soil into plant tissues is generally minimal; where high residues have been detected these are most likely to have arisen from atmospheric deposition or surface contamination by dust and soil particles. We *recommend* that when vegetables are grown in soils that are known to contain elevated lead levels, consumers should be advised to remove the outer leaves and wash the vegetables carefully before cooking and eating.

36

16.   We note that, as yet, a decline in lead concentrations in vegetables purchased at retail outlets has not been observed following the reduction in permitted amounts of lead added to petrol at the end of 1985. We *recommend* that monitoring should continue to investigate the effects of falling levels of lead in air.

February 1989

APPENDIX II CONSIDERATION OF THE REPORT BY THE FOOD ADVISORY COMMITTEE

1. We have been requested by the Steering Group on Food Surveillance to comment on this report. The Committee *welcomes* the reaction to their recommendations in the previous report[2] and in particular the further monitoring of diets and individual foods for lead. The Committee is encouraged by the steady diminution in lead intake from food and drink and the continued reduction in environmental sources of lead. Nevertheless efforts should continue to be made to reduce levels of lead in food and the environment. We wish to emphasise our support for the replacement of domestic lead water pipes since lead from tap-water especially in areas where the water is plumbosolvent contributes to overall lead intake. We *recommend* that monitoring of the diet should continue and that particular emphasis should be placed on establishing dietary lead intakes of young children, and women of child-bearing age.

2. The Committee *welcomes* the advances in canning technology made by the UK food industry. This has resulted in the elimination of lead solder from cans and in the reduction of lead levels in canned food. However, we note that some imported canned food is still contained in soldered cans. We *recommend* that efforts are made to encourage the use of welded cans worldwide. We note that the practice of storing food in opened cans may lead to elevated levels of metals in food and we endorse the recommendation of the Committee of Toxicity of Chemicals in Food, Consumer Products and the Environment (Appendix I) that more effective publicity be given to consumers to discourage this practice[29].

3. The Committee notes that consumption of wine from bottles with lead closures can result in elevated intakes of lead and we endorse the recommendation of the Committee on Toxicity of Chemicals in Food, Consumer Products and the Environment (Appendix I) that lead be substituted by an acceptable alternative for wine bottle seals. We also *recommend* that until lead capsules are replaced the situation is monitored and that consumers are regularly advised about contamination of wine from this source and in particular of the need to wipe the top and neck of the wine bottle before pouring[25]. We note that absorption of lead from beverages in humans is approximately twice that from food and *recommend* that further work be carried out on factors affecting absorption and on the forms of lead present in food and drink.

4. We *welcome* this report of the Steering Group and endorse the comments of the Committee on Toxicity. We wish to be kept informed of the results of future surveillance.

May 1989

APPENDIX III   COMMISSIONED PROJECTS WHOSE RESULTS ARE INCLUDED IN THIS REPORT

Title	Contractor(s)	Date	Total cost £K	Cost for lead component only	Published reference
AQA Scheme	LGC	1983	4	2	17
	AERE	1985	2.5	1.3	—
		1985–87	50+30+20	50+	—
Total diet study	BMRB	1982–87	351	56	—
	LGC				
Lead Levels in Individual Foods					
Selected ethnic foods	LGC	1981	5	3	—
Vegetables	ICLS }	1981	1	1	—
	CFPRA }				
	Avon County Scientific Services }				
	ADAS	1982	1	1	—
Wine and beer	AERE	1983	18	14	21
	AERE	1985	8	8	24
Individual foods	LGC	1983	28	5	—
Duplicate Diet Studies					
Control duplicate diet study	BMRB }	1982	30	21	—
	LGC }				
Duplicate diet study of Asian and Caucasian children	RSL }	1982	55	55	36
	Westminster Medical School }				

APPENDIX III *continued*

Title	Contractor(s)	Date	Total cost £K	Cost for lead component only	Published reference
Ayr duplicate diet study	BMRB LGC	1981	25	19	31, 43
Birmingham duplicate diet study	BMRB LGC	1984	46	35	37
Canned Food					
Monitoring	Avon County Scientific Services	1983–87	45	22.5	—
Storage of canned food	RSL	1983	5	5	28
Monitoring the Effect of the Reduction of Lead in Petrol					
Monitoring of indicator vegetables	BMRB LARS MAFF Food Science Labs	1983–87	90	90	—
Lead ratios in Italian wines	MAFF Food Science Labs	1986	2	2	—
Uptake of Lead by Plants					
Uptake of lead from soil	AERE	1982–83	50	50	—
Effect of preparation and cooking on lead in vegetables	Avon County Scientific Services	1982	13	4	—
Lead in urban garden soils and vegetables (to be completed)	Imperial College	1986–89	56	28	—

Title	Contractor(s)	Date	Total cost £K	Cost for lead component only	Published reference
Contribution made by Lead in Air					
Deposition of lead from air onto food	University of Nottingham	1980–83	28	28	—
To estimate the contribution of atmospheric lead to children's dietary lead	Westminster Medical School	1982–83	50	50	—
Absorption					
Absorption by man of lead from offal	AERE ARC, Compton	1981	10	10	51
Human lead absorption from vegetables	AERE	1982	15	15	
Uptake by man of lead in beer	AERE	1984	38	38	—
Absorption of lead by young children and infants	Westminster Medical School	1985–87	65	65	—
Human excretion of lead	AERE	1983	17	12	—
Lead concentrations in serial blood samples	University of Southampton	1982	1	1	41

Abbreviations

ADAS   Agricultural Development and Advisory Service
AERE   Atomic Energy Research Establishment, Harwell
BMRB   British Market Research Bureau
CFPRA   Campden Food Preservation Research Association
ICLS   International Consulting and Laboratory Service, Birmingham
LARS   Long Ashton Research Station
LGC   Laboratory of the Government Chemist
RSL   Research Services Ltd

41

APPENDIX IV   RESULTS OF ANALYSES

LIST OF TABLES

TABLE 1: **Lead concentrations in total diet samples and daily intake of lead 1982–87 (mg per kg fresh weight).**
Samples were obtained from retail outlets between 1982 and 1987. The Total Diet is made up of food groups representing the average diet consumed in Great Britain, as described in Reference 7.

Food group	Estimated weight of food eaten (kg/day)	Lead concentration (mg/kg) Mean (Range) 1982[a]	1983[a]	1984[a]	Estimated daily intake of lead (mg/person/day) 1982	1983	1984
Bread and cereals	0.241	<0.050 (<0.05–0.05)	<0.050 (<0.05–0.05)	<0.050 (<0.05–0.05)	0.012	0.012	0.012
Meat and poultry	0.056	<0.053 (<0.05–0.10)	<0.050 (<0.05–0.05)	<0.050 (<0.05–0.05)	0.003	0.003	0.003
Offal	0.003	<0.169 (<0.05–0.35)	<0.173 (<0.05–0.70)	<0.100 (<0.05–0.30)	0.001	0.001	0.000
Meat products	0.075	<0.059 (<0.05–0.15)	<0.050 (<0.05–0.05)	<0.056 (<0.05–0.15)	0.004	0.004	0.004
Fish	0.017	<0.084 (<0.05–0.20)	<0.157 (<0.01–0.90)	<0.121 (<0.05–0.60)	0.001	0.003	0.002
Oils	0.094	<0.053 (<0.05–0.10)	<0.050 (<0.05–0.05)	<0.050 (<0.05–0.05)	0.005	0.005	0.005
Sugars and preserves	0.090	<0.050 (<0.05–0.05)	<0.050 (<0.05–0.05)	<0.050 (<0.05–0.05)	0.005	0.005	0.005
Green vegetables	0.047	<0.053 (<0.05–0.10)	<0.050 (<0.05–0.05)	<0.050 (<0.05–0.05)	0.002	0.002	0.002
Potatoes	0.156	<0.020 (<0.02–0.02)	<0.020 (<0.02–0.02)	<0.025 (<0.02–0.06)	0.003	0.003	0.004
Other vegetables	0.070	<0.053 (<0.05–0.10)	<0.050 (<0.05–0.05)	<0.056 (<0.05–0.10)	0.004	0.004	0.004
Canned vegetables	0.042	0.169 (0.05–0.30)	0.153 (0.05–0.45)	0.130 (0.05–0.20)	0.007	0.006	0.004
Fruit	0.061	<0.050 (<0.05–0.05)	<0.050 (<0.05–0.05)	<0.050 (<0.05–0.05)	0.003	0.003	0.003
Fruit products	0.030	<0.172 (<0.05–0.30)	<0.140 (<0.05–0.60)	0.112 (0.05–0.25)	0.005	0.004	0.003
Beverages	0.665	<0.010 (<0.01–0.01)	<0.010 (<0.01–0.01)	<0.011 (<0.01–0.02)	0.007[b]	0.007[b]	0.007[b]
Milk	0.335	<0.020 (<0.02–0.02)	All <0.02	All <0.02	0.007	0.007	0.007

43

TABLE 1 *continued*

Food group	Estimated weight of food eaten (kg/day)	Lead concentration (mg/kg) Mean (Range)			Estimated daily intake of lead (mg/person/day)		
		1985[a]	1986[a]	1987[a]	1985	1986	1987
Bread and cereals	0.241	<0.050 (<0.05–0.05)	<0.050 (<0.05–0.05)	<0.050 (<0.05–0.05)	0.012	0.012	0.012
Meat and poultry	0.056	All <0.05	<0.050 (<0.05–0.05)	<0.050 (<0.05–0.05)	0.003	0.003	0.003
Offal	0.003	<0.112 (<0.05–0.35)	<0.123 (<0.05–0.35)	<0.084 (<0.05–0.20)	0.000	0.000	0.001
Meat products	0.075	<0.056 (<0.05–0.10)	<0.050 (<0.05–0.05)	<0.066 (<0.05–0.15)	0.004	0.004	0.005
Fish	0.017	<0.097 (<0.05–0.40)	<0.057 (<0.05–0.10)	<0.068 (<0.05–0.15)	0.002	0.001	0.001
Oils	0.094	<0.050 (<0.05–0.05)	<0.050 (<0.05–0.05)	<0.050 (<0.05–0.05)	0.005	0.005	0.004
Sugars and preserves	0.090	<0.056 (<0.05–0.10)	<0.050 (<0.05–0.05)	<0.050 (<0.05–0.05)	0.005	0.004	0.004
Green vegetables	0.047	<0.050 (<0.05–0.05)	<0.050 (<0.05–0.05)	<0.050 (<0.05–0.05)	0.002	0.002	0.002
Potatoes	0.156	<0.020 (<0.02–0.02)	<0.020 (<0.02–0.02)	<0.020 (<0.02–0.02)	0.003	0.003	0.003
Other vegetables	0.070	<0.050 (<0.05–0.05)	<0.050 (<0.05–0.05)	<0.050 (<0.05–0.05)	0.003	0.003	0.004
Canned vegetables	0.042	<0.135 (<0.05–0.05)	<0.070 (<0.05–0.25)	<0.050 (<0.05–0.05)	0.006	0.003	0.002
Fruit	0.061	<0.050 (<0.05–0.05)	<0.050 (<0.05–0.05)	<0.050 (<0.05–0.05)	0.003	0.003	0.003
Fruit products	0.030	<0.132 (<0.05–0.25)	<0.100 (<0.05–0.20)	<0.058 (<0.05–0.10)	0.004	0.003	0.002
Beverages	0.665	<0.010 (<0.01–0.01)	<0.010 (<0.01–0.01)	<0.010 (<0.01–0.01)	0.007[b]	0.007[b]	0.011[b]
Milk	0.335	All <0.02	All <0.02	All <0.02	0.007	0.007	0.006

Notes:—a: Number of diet samples analysed each year was as follows:
16 in 1982; 15 in 1983; 17 in 1984; 17 in 1985; 15 in 1986; 19 in 1987.

b: Corrected for dilution as indicated in para.16.

44

TABLE 2: Lead in individual foodstuffs (mg/kg fresh weight)

Food	Number of samples	Mean (mg/kg)	Range (mg/kg)
Cereals			
Flour			
Self-raising	8	<0.05	<0.05–0.05
Plain	8	<0.05	—
Wheatmeal	3	<0.05	—
Wholemeal	5	<0.09	<0.05–0.15
Bread			
White	8	<0.05	<0.05–0.05
Wholemeal	8	<0.05	—
Plain biscuits	5	<0.05	—
Rice	5	<0.06	<0.05–0.10
Meat			
Beef (braising steak)	8	<0.05	<0.05–0.05
Lamb	8	<0.05	<0.05–0.10
Pork	8	<0.05	<0.05–0.05
Chicken	10	<0.05	—
Concentrated Soft Drinks			
Orange	6	<0.01	<0.01–0.01
Lemon	2	<0.01	<0.01–0.01
Lime	2	<0.01	<0.01–0.01
Lemon and lime	1	<0.01	—
Blackcurrant	2	0.01	—
Raspberry	1	<0.01	—
Other Foods			
Baby foods			
cereal and dried	9	<0.07	<0.05–0.15
in jars	11	<0.05	<0.05–0.05
Chutney	5	<0.08	<0.05–0.20
Dried milk	10	<0.06	<0.05–0.10
Dried soup	5	<0.05	<0.05–0.05
Fish pastes	8	<0.06	<0.05–0.15
Stock cubes	5	<0.12	<0.05–0.35
Frozen Foods			
Carrots	10	<0.05	<0.05–0.05
Spinach	11	<0.05	<0.05–0.05
Strawberries	4	<0.03	<0.02–0.06
Raspberries	4	<0.02	<0.02–0.02
Blackcurrants	4	0.05	0.04–0.06
Blackberries	4	0.04	0.02–0.08

TABLE 2 *continued*

Food	Number of samples	Mean (mg/kg)	Range (mg/kg)
Fresh Fruit			
Pears	8	<0.03	<0.02–0.06
Apples	8	0.03	0.02–0.06
Tomatoes	2	<0.02	—
Gooseberries	1	0.06	—
Loganberries	1	0.16	—
Blackberries	4	0.07	0.02–0.14

TABLE 3: Lead in selected ethnic foodstuffs (mg/kg fresh weight)

Food	Number of samples	Mean (mg/kg)	Range (mg/kg)
Chinese Foods			
Salted egg	1	0.25	—
Fishball	1	0.05	—
Salted fish	1	0.25	—
Chinese sausage	1	0.05	—
Chinese luncheon meat	1	0.40	—
Dried cole	1	0.15	—
Soup mix	1	<0.05	—
Oyster sauce	1	0.10	—
Bean curd	1	<0.05	—
Soya bean drink	1	0.05	—
Rice wine	1	0.07	—
Pastries			
Sweet	3	0.13	0.05–0.20
Savoury	1	0.05	—
Brown rice	1	0.05	—
Canned Foods			
Kantola	1	0.40	—
Karela	1	0.50	—
Lilva	1	0.20	—
Okra	1	1.40	—
Parval	1	0.45	—
Tinda	1	0.35	—
Jackfruit	1	<0.05	—
Breadfruit	1	0.10	—
Balor beans	1	0.15	—
Papri beans	1	0.20	—
Anchovies	1	0.90	—
Fresh Vegetables and Beans			
Green banana	1	0.05	—
Gherkins	1	<0.05	—
Karela	1	<0.05	—
Parval	1	0.10	—
Patra leaves	1	0.10	—
Turia	1	<0.05	—
Vine leaves	1	0.15	—
Balor beans	1	<0.05	—
Guare beans	1	<0.05	—
Papri beans	1	0.05	—
Other Foods			
Ghee	3	0.15	0.05–0.30
Fetta cheese	1	0.05	—
Pork, dried	1	0.10	—
Fish			
Dried	3	0.18	0.10–0.30
Frozen	2	<0.05	<0.05–0.0

TABLE 3 *continued*

Food	Number of samples	Mean (mg/kg)	Range (mg/kg)
Taramasalata	1	<0.05	—
Asian pickle	4	0.95	0.50–1.70
Hummus	1	0.10	—
Asian sweets	5	<0.27	<0.05–1.00
Jaggery	1	<0.05	—
Turkish Delight	1	0.05	—
Asian pastries	1	0.05	—
Greek pastries	1	0.05	—
Popadums	1	1.00	—
Pitta bread	1	<0.05	—
West Indian bread	1	<0.05	—

TABLE 4: Lead in fresh vegetables (mg per kg fresh weight)

Vegetable	Number of samples	Mean lead content (mg/kg)		Region of purchase			
				Bristol	Dawlish	Leeds	London
Sampled in November 1982							
Potatoes[a]	12	0.03	Mean	0.06	<0.02	<0.03	<0.03
			Range	0.04–0.10	<0.02–0.02	<0.02–0.04	<0.02–0.03
			N[b]	3	3	3	3
Carrots[a]	16	0.04	Mean	0.03	<0.03	0.09	<0.02
			Range	0.02–0.05	<0.01–0.05	0.04–0.13	<0.01–0.03
			N[b]	4	4	4	4
Sprouts	8	0.03	Mean	0.04	<0.02	<0.03	0.04
			Range	0.03–0.05	<0.02–0.02	<0.02–0.03	—
			N[b]	2	2	2	2
Leeks	8	0.05	Mean	0.05	0.04	0.04	0.08
			Range	0.01–0.09	0.03–0.04	—	0.04–0.12
			N[b]	2	2	2	—
Spinach	9	0.26	Mean	0.21	0.27	NA[c]	0.31
			Range	0.03–0.32	0.09–0.42	—	0.11–0.51
			N[b]	3	4	—	2
Cabbage	8	0.06	Mean	0.13	0.02	0.02	<0.07
			Range	0.09–0.16	0.01–0.03	—	<0.01–0.13
			N[b]	2	2	2	—

TABLE 4 *continued*

Vegetable	Number of samples	Mean lead content (mg/kg)		Region of purchase			
				Bristol	Dawlish	Leeds	London
Sampled in March 1983							
Potatoes[a]	9	0.04	Mean	<0.03	<0.02	<0.06	0.02
			Range	<0.02–0.03	—	<0.03–0.11	0.02–0.025
			N[b]	3	—	3	2
Cabbage	8	0.07	Mean	<0.02	0.05	<0.19	0.02
			Range	<0.02–0.02	0.04–0.07	<0.02–0.35	0.02–0.03
			N[b]	2	2	2	2
Sprouts	6	0.07	Mean	0.07	<0.02	0.07	0.08
			Range	0.05–0.09	—	0.05–0.08	0.06–0.09
			N[b]	2	—	2	2
Leeks	8	0.10	Mean	0.20	0.04	0.06	0.09
			Range	0.04–0.36	—	0.04–0.09	0.02–0.16
			N[b]	2	2	2	—

Notes: a: Samples washed to remove obvious soil contamination.
 b: Numbers of samples.
 c: Not analysed.

50

TABLE 5: Lead in edible bone proteins, bonemeal tablets and dolomite tablets (mg/kg fresh weight)

Product	Number of samples	Mean (mg/kg)	Range (mg/kg)
Bone protein	4	0.08	0.05–0.10
Bone collagen	4	0.21	0.20–0.25
Bone phosphate	4	3.90	2.70–4.40
Bonemeal tablets	20	4.40	1.30–9.40
Dolomite tablets	4	2.30	0.80–5.30

TABLE 6: Concentration of lead found in mussels sampled in 1985 (mg/kg fresh weight)

Area	Number of shellfish in bulked samples	Mean length (cm)	Concentration[a]
Tyne	51	4.4	2.9
Tees	102	4.6	4.9
Humber	120	4.1	<0.6
Wash	94	4.6	<0.6
Thames	100	4.4	0.7
Eastern Channel	150	4.4	0.8
Western Channel	148	4.1	2.7
Severn	75	3.6	1.5
Cardigan Bay	100	4.0	1.6
Liverpool Bay	84	4.6	1.1
Morecambe Bay	100	4.4	1.2
Solway	100	4.6	1.4

Note: a: Results of duplicate analyses on bulked samples.

TABLE 7: Lead concentrations in canned foodstuffs (mg per kg fresh weight)

Food	Autumn 1983		Spring 1984		Autumn 1984		Spring 1985	
	N	Mean[a] (Range)	N	Mean[a] (Range)	N	Mean[a] (Range)	N	Mean[a] (Range)
Asparagus	6	0.06 (<0.10–0.18)	4	0.14 (<0.10–0.35)	6	0.17 (<0.10–0.34)	6	0.11 (<0.10–0.15)
Blackcurrants	4	0.94 (<0.10–3.30)	3	0.48 (0.12–1.00)	5	0.26 (<0.10–0.55)	4	0.28 (<0.20–0.39)
Pineapple	6	0.19 (0.10–0.40)	6	0.09 (<0.10–0.17)	6	0.23 (0.12–0.48)	6	0.34 (0.20–0.55)
Sardines	6	1.61 (0.23–3.00)	6	1.95 (<0.10–2.90)	7	1.24 (<0.10–2.40)	10	1.74 (0.12–5.40)
Tomato soup	6	0.02 (<0.10–0.10)	6	0.02 (<0.10–0.13)	6	0.08 (<0.10–0.30)	6	0.17 (<0.10–0.35)
Spaghetti	6	0.16 (<0.10–0.30)	6	0.08 (<0.10–0.27)	6	0.26 (<0.10–1.20)	6	0.10 (<0.10–0.17)
Baked beans	6	0.07 (<0.10–0.18)	6	0.17 (<0.10–0.65)	6	0.10 (<0.10–0.15)	6	0.07 (<0.10–0.42)
Pears	6	0.15 (<0.10–0.60)	6	0.13 (<0.10–0.40)	6	0.29 (0.12–0.48)	6	0.35 (<0.15–0.47)
Carrots	6	0.17 (<0.10–0.68)	6	0.20 (<0.10–1.20)	6	0.04 (<0.10–0.12)	6	0.28 (0.15–0.65)
Tomatoes	6	0.31 (<0.10–2.30)	6	0.15 (0.10–0.32)	6	0.16 (0.10–0.52)	6	0.25 (0.15–0.44)
Blackberries	2	1.00 (—)	3	1.24 (0.45–2.30)	3	2.71 (0.62–6.10)	3	0.61 (0.20–1.32)
Evaporated milk	6	0.03 (<0.10–0.20)	6	0.05 (<0.10–0.15)	6	<0.10 (—)	0	NA[b]
Fruit cocktail	6	0.45 (<0.10–1.75)	6	0.11 (<0.10–0.29)	6	0.27 (<0.10–0.56)	7	0.39 (0.17–0.67)
Red plums	5	0.36 (<0.10–0.90)	5	0.16 (<0.10–0.31)	5	0.17 (<0.10–0.35)	5	0.36 (0.27–0.45)
Corned beef	6	0.41 (0.20–0.73)	7	0.60 (<0.22–1.50)	6	0.80 (0.24–2.90)	6	0.65 (0.17–2.23)
Cream	6	0.03 (<0.10–0.15)	6	0.02 (<0.10–0.10)	6	0.06 (<0.10–0.34)	7	0.19 (0.10–0.32)
Rhubarb	6	0.73 (0.33–0.98)	6	0.54 (0.32–0.81)	0	NA[b]	6	0.27 (0.17–0.42)
Apricots	6	0.25 (0.10–0.38)	6	0.22 (<0.10–0.34)	6	0.26 (0.14–0.42)	5	0.32 (0.25–0.42)

Notes: a: Means calculated by taking all measurements reported as <x as 0.
b: Not analysed.

52

TABLE 7 *continued*

Food	Autumn 1985 N	Mean[a] (Range)	Spring 1986 N	Mean[a] (Range)	Autumn 1986 N	Mean[a] (Range)	Spring 1987 N	Mean[a] (Range)
Asparagus	9	0.02 (<0.10–0.10)	6	0.11 (<0.11–0.47)	5	0.19 (<0.10–0.50)	5	0.10 (—)
Blackcurrants	2	0.10 (<0.10–0.19)	2	<0.10(—)	3	<0.10(—)	3	0.23 (<0.10–0.69)
Pineapple	6	0.04 (<0.10–0.13)	6	0.14 (<0.12–0.83)	6	<0.10(—)	6	0.10 (<0.10–0.31)
Sardines	26	0.99 (<0.10–2.54)	19	0.17 (0.11–1.59)	13	0.37 (<0.10–2.10)	12	0.18 (<0.10–0.36)
Tomato soup	6	<0.10(—)	6	<0.10(—)	6	<0.10(—)	6	<0.10(—)
Spaghetti	6	<0.10(—)	6	<0.10(—)	6	<0.10(—)	5	<0.10(—)
Baked beans	6	<0.10(—)	6	<0.10(—)	6	<0.10(—)	7	<0.10(—)
Pears	6	<0.10(—)	6	0.09 (<0.10–0.37)	6	0.17 (<0.10–0.56)	6	0.03 (<0.10–0.16)
Carrots	6	0.14 (<0.10–0.86)	6	0.11 (<0.10–0.53)	6	0.09 (<0.10–0.46)	6	0.08 (<0.10–0.45)
Tomatoes	7	0.17 (<0.10–0.36)	6	0.14 (<0.11–0.39)	7	0.07 (<0.10–0.19)	6	0.09 (<0.10–0.55)
Blackberries	4	0.42 (0.10–1.26)	4	0.04 (<0.12–0.17)	3	<0.10(—)	3	<0.10(—)
Evaporated milk	0	NA[b]	0	NA[b]	0	NA[b]	0	NA[b]
Fruit cocktail	6	0.09 (<0.10–0.31)	6	0.04 (<0.10–0.23)	8	0.10 (<0.10–0.21)	6	0.05 (<0.10–0.20)
Red plums	7	<0.10(—)	6	<0.10(—)	5	0.04 (<0.10–0.21)	7	<0.10(—)
Corned beef	8	0.73 (0.11–1.83)	8	0.72 (0.12–1.79)	7	0.91 (<0.10–3.10)	6	0.42 (0.11–0.75)
Cream	6	0.03 (<0.10–0.16)	5	0.02 (<0.10–0.12)	6	<0.10(—)	6	<0.10(—)
Rhubarb	6	0.05 (<0.10–0.30)	6	<0.10(—)	6	<0.10(—)	6	0.09 (<0.10–0.37)
Apricots	7	0.22 (<0.10–0.38)	6	0.26 (0.12–0.31)	6	0.10 (<0.10–0.30)	7	0.17 (<0.10–0.23)

Notes: a: Means calculated by taking all measurements reported as <x as 0.
b: Not analysed.

E

53

TABLE 8: The effect of preparation and cooking on the lead content of fresh vegetables (mg/kg fresh weight)

Vegetable	Number of Samples	Treatment[a]			
		A	B	C	D
Cabbage	10	1.1	0.7[b]	0.4	0.3
Carrots	10	3.5	1.9	0.3	0.1
Leeks	10	3.2	0.5	0.2	0.1
Lettuce	10	4.7	1.4	0.5	0.5
Sprouts	10	0.4	0.2	0.1	0.1

Notes: a: Treatment A: vegetables washed but untrimmed.
Treatment B: vegetables trimmed but unwashed.
Treatment C: vegetables washed, trimmed and cut up.
Treatment D: vegetables washed, trimmed, cut up, boiled and drained as if for consumption.
b: One sample not analysed because of soil contamination.

TABLE 9: Comparison of estimated lead intakes in the UK with those for other countries reported in the literature[a]

Country	Duplicate Diet		Total Diet	
	Intake (μg/day)	Reference	Intake (μg/day)	Reference
France	108	53	—	—
West Germany	58	54	—	—
Netherlands	107	55	—	—
India	550	56	—	—
Italy	108	57	210	60
Belgium	179	58	290	61
Austria	66	59	210	59
UK	44	(This Report)	20–60	(This Report)
Denmark	—	—	80	62
Finland	—	—	70	63
USA	—	—	60	64

Note: a: Care is required when interpreting these data due to the differences in sampling, analytical methodology and methods of estimating intakes used by different countries.

APPENDIX V ANALYTICAL QUALITY ASSURANCE

Initial Scheme

1.   The first intercomparison exercise was conducted between September 1982 and June 1983 to assess the analytical performance of non-government laboratories who regularly analysed foodstuffs for heavy metals, including lead and cadmium. Twenty-eight participating non-government laboratories analysed 6 dried food samples and 2 National Bureau of Standards (NBS) samples for lead and cadmium. Laboratories were asked to analyse the 6 food samples (2 livers, 2 fishmeals, cabbage and curly kale) in duplicate for both lead and cadmium on 3 occasions at intervals of 1 month, giving a total of 36 samples and 72 analyses. The samples had previously been tested for homogeneity and analysed by government laboratories who were asked to make further duplicate analyses in the second round of this survey.

Further Main Scheme

2.   Following the preparation and distribution of a questionnaire designed to assess the methods and procedures used in the participating laboratories, the Working Party's AQA Sub-Group implemented a sequential programme of work which involved the preparation, distribution and analysis of specially prepared samples which increased in analytical complexity with each stage of the programme. The 6 stages of this programme began with the distribution of standard solutions in July 1985 and culminated in June/July 1987 with the analyses of homogeneous samples based on milk, cabbage, and starch (both spiked and unspiked). Intermediate stages included samples based on starch powder which were spiked with major cations and cadmium and lead, and unspiked and spiked liver and cabbage powders. During the scheme 1 set of identical samples was incorporated into an FAO analytical intercomparison exercise involving laboratories world-wide.

APPENDIX VI THE EFFECTS OF THE REDUCTION OF LEAD IN PETROL ON THE LEAD CONTENT OF GREEN VEGETABLES

1. To minimise human exposure to lead the amount added to petrol was reduced during 1985, and from January 1986[12] the maximum permitted level was 0.15 g/l compared with the previous limit of 0.4 g/l. To assess the effectiveness of this measure a number of studies were initiated to monitor the expected decrease of lead in foodstuffs and the environment. As part of this programme a 5 year study was implemented to measure any reduction in the lead content of green vegetables.

2. Some 520 samples of cabbage, Brussels sprouts, kale, spring greens and lettuce were purchased at retail outlets in 60 towns in the UK between September 1983 and April 1988. The large majority of these were obtained at greengrocers and market stalls. Due to problems of local availability it was necessary to obtain some 60 per cent of the kale samples at supermarket chains.

3. Details of the analytical procedure employed in the study are given elsewhere[83]. The method was validated by analysis of standard reference materials. These materials, together with in-house reference samples and 'field blanks', were regularly analysed during the course of the survey to ensure the accuracy and reliability of the analytical data. Lettuce samples were initially washed prior to chopping, drying and analysis to minimise the effects of soil contamination. Kale, spring greens, Brussels sprouts and the majority of the cabbage samples were not washed as it was considered that this might remove some of the petrol-derived lead. As these vegetables are however washed under normal domestic circumstances, approximately 57 per cent of the cabbage samples were divided into 2 halves and one half washed to assess the effects on the lead concentration.

4. All samples were analysed at the MAFF Food Science Laboratory (Norwich). From September 1983 to January 1986, the samples were prepared at the Long Ashton Research Station (Bristol). Due to a reorganisation of the Agricultural and Food Research Council it was not possible for preparation of the vegetables to continue at this location. As a result this function was transferred to MAFF Food Science Laboratory in Norwich where it was performed until the end of the study in April 1988. To check whether this change in the site of sample preparation influenced the lead content of the vegetables a period of overlap was arranged in the spring of 1986. During this time double the quantities of Brussels sprouts and cabbage were purchased, divided into 2 and separately prepared at the 2 sites.

5. The results obtained for the 659 analytical samples from the 524 vegetables purchased over the 5 year period are summarised in Table A and shown in more detail for each class of vegetable in Tables B to G. The data are also illustrated

in the form of histograms in Figure A. Only one kale value was available for spring 1986 and the resulting mean bar was not shown. The concentrations are in units of µg/kg and are all expressed on a dry weight basis.

6.   Comparison of the lead content of the Brussels sprouts prepared at Long Ashton and Norwich during the overlap period in the spring of 1986 revealed that there was no significant difference (Student's paired t-test) between the sites. In contrast, the results for cabbages prepared at Norwich were significantly lower ($p < 0.05$). The average values for the washed and unwashed samples prepared at this location were respectively 11 µg/kg and 10µg/kg less than those processed at Long Ashton. To compensate for this effect the results for the cabbage samples prepared at Norwich have all been increased by 10 or 11 µg/kg as appropriate. These modified values have been incorporated in Figure A and were employed in the subsequent statistical analyses. The summary data in Table A show that the mean lead content of the Brussels sprouts (310 µg/kg) is some 5 times higher than the mean content of the cabbage samples. It is probable that the comparatively small mean difference of 10 µg/kg observed for the cabbage samples would not have been significant for the Brussels sprouts due to the higher lead levels and correspondingly larger variability involved. The same would also be true for the samples of kale, spring greens and lettuce as these have even higher levels of lead.

7.   The data in Figure A and Tables A to G show that there is considerable variability in the lead levels found for each class of vegetable. To assess the influence of the various factors on these results the data were examined in the first instance by analysis of variance. In the case of kale and Brussels sprouts there was a significant seasonal factor with spring crops containing higher concentrations than the corresponding autumn samples ($p < 0.001$). This trend was presumably a consequence of the longer growing season of the spring harvested vegetables. Analysis of variance also revealed that there was no significant difference between years during 1983 to 1988 for cabbage, lettuce, kale or Brussels sprouts. There was however a significant difference between years for spring greens ($p < 0.025$). This statistical test was repeated after the removal of outliers (Dixon's test) from each growing season for each vegetable. This showed that there was a significant difference between years for Brussels sprouts ($p < 0.01$) after outlier removal. There was however still no significant difference between years for kale, cabbage and lettuce. In the case of spring greens there was no significant difference between years after outlier removal in contrast to the earlier finding when outliers were included. The spring green sample outliers were principally associated with high results in samples obtained from Plymouth in 1985. The reasons for these elevated results are unknown but they may have biased the 1985 results as samples from this location were not included in the other 3 years. In view of this likely bias it seems reasonable to exclude these samples as outliers in which case there is no significant between years effect for the spring greens.

57

8.   The results were also analysed by a one-sided t-test to specifically ascertain whether the mean lead concentrations were significantly lower after January 1986, i.e. following the reduction in petrol lead. Each of the 5 vegetable classes were pooled into essentially 2 groups, namely pre- and post-January 1986. Where appropriate the groups were subdivided to take account of significant factors such as spring/autumn and outliers removed. Additionally samples taken during January to March 1986 were not included as this was a period of transition. Employing this more powerful technique it was found that there was a significant decrease after January 1986 for spring ($p < 0.025$) and autumn ($p < 0.05$) Brussels sprouts, spring greens ($p < 0.001$) and spring kale ($p < 0.025$). There was however no significant decrease for cabbage, lettuce or autumn kale. It is quite possible that the statistically significant reductions in the mean lead content of the Brussels sprouts, spring greens and spring kale purchased after January 1986 may at least in part have been due to the decrease in petrol lead. However, the wide range of lead concentrations encountered illustrate that a number of factors can affect the lead levels in vegetables. The possibility that one or more unidentified factors were responsible for the observed changes cannot be excluded and in this context it should be noted that no similar temporal trend was observed for the cabbage, lettuce and autumn kale.

9.   The general absence of a major reduction in lead levels for all classes of the retail vegetables since the beginning of 1986 suggests that lead recently emitted from automobile exhausts is only one of a number of factors that influence the lead content of such foodstuffs. Other important parameters are likely to include the lead content of the soil and the presence of environmental lead arising from automobile exhaust emitted prior to 1986. As far as this historic automobile exhaust lead is concerned it is to be expected that the recent reduction in petrol lead will have a cumulative rather than immediate effect in decreasing contamination from this source. It is well established[65] that recently emitted automobile exhaust makes a significantly greater contribution to environmental lead contamination in urban as compared to rural areas. It is therefore perhaps not surprising that whilst there has been a recent decrease in blood lead levels in urban areas in the UK[84] a similar effect has only been observed for some of the retail green vegetable classes examined in this study. Further longer term monitoring would be required to investigate the cumulative effects of the reduction in petrol lead on its presence in green vegetables.

TABLE A: Summary of the lead concentrations found in 'indicator' vegetables (μg per kg dry weight)

Vegetable	Sample number	Mean	Standard error of mean	Range
Washed cabbage	80	45	3	10–190
Cabbage	140	67	3	15–310
Brussels sprouts	120	310	15	73–1,300
Lettuce	118[a]	897[a]	80[a]	170–5,000[a]
Spring greens	59	1,090	112	20–4,000
Curly kale	85	2,900	178	590–8,400

Note: a: For lettuce two very high results were found which were an order of magnitude larger than the remainder. These results have been excluded from the calculation of summary statistics.

TABLE B: Lead in indicator samples of cabbage (washed) (μg per kg dry weight)

Season	Prepara-tion site[a]	Sample number	Mean	Standard error of mean	Lowest result	Highest result	Second highest result	Third highest result
Spring 1984	LA	20	44	3	21	79	65	61
Spring 1985	LA	20	58	8	22	190	93	84
Spring 1986	LA	20[b]	46	4	26	79	77	68
	N	20[b]	35	3	20	61	60	55
Spring 1987	N	20	40	7	10	130	94	68

Note: a: LA = Long Ashton Research Station.
N = MAFF Food Science Laboratory (Norwich).

b: Divided samples.

TABLE C: Lead in indicator samples of cabbage (unwashed) (μg per kg dry weight)

Season	Preparation site[a]	Sample number	Mean	Standard error of mean	Lowest result	Highest result	Second highest result	Third highest result
Autumn 1983	LA	15	150	21	59	310	280	260
Spring 1984	LA	20	56	5	28	96	93	88
Autumn 1984	LA	15	64	6	21	96	91	82
Spring 1985	LA	20	64	5	35	120	100	99
Autumn 1985	LA	15	64	8	36	150	110	81
Spring 1986	LA	20[b]	59	6	28	140	91	86
	N	20[b]	49	5	15	98	91	79
Autumn 1986	N	15	72	9	35	160	110	110
Spring 1987	N	20	42	4	21	77	71	60

Note: a: LA = Long Ashton Research Station.
  N = MAFF Food Science Laboratory (Norwich).
  b: Divided samples.

TABLE D: Lead in indicator samples of Brussels sprouts (μg per kg dry weight)

Season	Prepara-tion site[a]	Sample number	Mean	Standard error of mean	Lowest result	Highest result	Second highest result	Third highest result
Autumn 1983	LA	15	255	26	113	510	390	290
Spring 1984	LA	15	459	66	184	1,300	630	570
Autumn 1984	LA	15	245	40	73	540	520	420
Spring 1985	LA	15	414	55	179	970	610	580
Autumn 1985	LA	15	208	31	98	560	400	240
Spring 1986	LA	15[b]	499	48	180	780	680	680
	N	15[b]	500	56	220	920	810	750
Autumn 1986	N	15	162	23	85	410	220	220
Spring 1987	N	15	329	49	130	870	520	500

Note: a: LA = Long Ashton Research Station.
       N = MAFF Food Science Laboratory (Norwich).

       b: Divided samples.

62

TABLE E: Lead in indicator samples of lettuce (washed) (μg per kg dry weight)

Season	Preparation site[a]	Sample[b] number	Mean[b]	Standard[b] error of mean	Lowest result	Highest result	Second highest result	Third highest result
Autumn 1983	LA	3	891	180	540	1,090	1,070	—
Spring 1984	LA	26	1,020	180	300	(37,000)	3,800	3,500
Autumn 1984	LA	15	697	130	320	2,300	1,060	890
Spring 1985	LA	15	758	130	330	2,300	1,200	1,100
Autumn 1985	LA	15	1,280	300	330	5,000	2,100	1,700
Spring 1986	LA / N	3 } / 12 }	515	83	170	1,200	1,100	850
Autumn 1986	N	14	970	290	180	(31,000)	3,700	3,100
Spring 1987	N	15	955	300	230	4,600	2,500	1,400

Note: a: LA = Long Ashton Research Station.
          N = MAFF Food Science Laboratory (Norwich).

         b: For lettuce two very high results, shown in brackets, were found which were an order of magnitude larger than the remainder. These results have
             been excluded from the calculation of summary statistics.

63

TABLE F: Lead in indicator samples of spring greens (µg per kg dry weight)

Season	Prepara-tion site[a]	Sample number	Mean	Standard error of mean	Lowest result	Highest result	Second highest result	Third highest result
Spring 1984	LA	15	965	150	280	2,300	1,600	1,400
Spring 1985	LA	16	1,650	320	410	4,000	3,900	3,800
Spring 1986	LA N	3 } 11 }	1,070	230	200	3,700	1,400	1,400
Spring 1987	N	14	584	72	260	1,100	1,000	930

Note: a: LA = Long Ashton Research Station.
N = MAFF Food Science Laboratory (Norwich).

TABLE G: Lead in indicator samples of curly kale (μg per kg dry weight)

Season	Preparation site[a]	Sample number	Mean	Standard error of mean	Lowest result	Highest result	Second highest result	Third highest result
Autumn 1983	LA	8	2,920	510	1,400	5,100	4,800	3,600
Spring 1984	LA	5	4,420	770	3,300	7,300	4,600	3,500
Autumn 1984	LA	15	2,080	260	810	4,100	4,000	2,600
Spring 1985	LA	7	4,870	600	2,800	7,700	6,000	5,000
Autumn 1985	LA	15	2,670	470	810	5,700	5,400	4,900
Spring 1986	N	1	930	—	—	—	—	—
Autumn 1986	N	9	1,870	420	590	4,800	2,700	2,000
Spring 1987	N	11	3,980	790	900	8,400	6,900	6,200
Autumn 1987	N	6	2,480	620	1,000	4,800	3,800	2,200
Spring 1988	N	8	2,380	430	1,100	5,200	2,500	2,400

Note: a: LA = Long Ashton Research Station.
N = MAFF Food Science Laboratory (Norwich).

FIGURE A: Mean lead concentrations (μg/kg dry weight) found in selected indicator vegetables in either autumn or spring 1983–1987

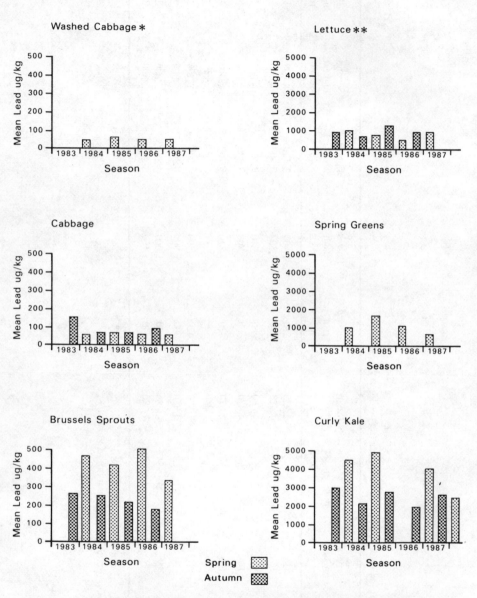

*Cabbage and Washed Cabbage data from samples prepared at the Norwich based MAFF Food Science Laboratory have been increased by 10 and 11 ug/kg respectively to allow for the difference due to the site prepartion

**The data shown have had the two very high results in excess of 30000 ug/kg removed